ACS SYMPOSIUM SERIES **410**

Historic Textile and Paper Materials II

Conservation and Characterization

S. Haig Zeronian, EDITOR
University of California—Davis

Howard L. Needles, EDITOR
University of California—Davis

Developed from a symposium sponsored
by the Cellulose, Paper, and Textile Division
at the 196th National Meeting
of the American Chemical Society,
Los Angeles, California,
September 25–30, 1988

American Chemical Society, Washington, DC 1989

Library of Congress Cataloging-in-Publication Data

Historic textile and paper materials II: conservation and characterization
    S. Haig Zeronian, editor,   Howard L. Needles, editor

    p.   cm.—(ACS Symposium Series, 0097–6156; 410)

    "Developed from a symposium sponsored by the
Cellulose, Paper, and Textile Division at the 196th National
Meeting of the American Chemical Society, Los Angeles,
California, September 25–30, 1988.

    Includes bibliographical references.

    ISBN 0–8412–1683–5
    1. Textile fabrics—Conservation and restoration—
Congresses. 2. Paper—Preservation—Congresses.

    I. Zeronian, S. Haig, 1932–   . II. Needles, Howard L.
III. American Chemical Society. Cellulose, Paper, and
Textile Division. IV. American Chemical Society. Meeting
(196th: 1988: Los Angeles, Calif.). V. Series

TS1449.H57   1989
746—dc20   89–38410
                                         CIP

The paper used in this publication meets the minimum requirements of American National Standard for Information Sciences—Permanence of Paper for Printed Library Materials, ANSI Z39.48–1984.

ACS Symposium Series

M. Joan Comstock, *Series Editor*

1989 ACS Books Advisory Board

Foreword

The ACS SYMPOSIUM SERIES was founded in 1974 to provide a medium for publishing symposia quickly in book form. The format of the Series parallels that of the continuing ADVANCES IN CHEMISTRY SERIES except that, in order to save time, the papers are not typeset but are reproduced as they are submitted by the authors in camera-ready form. Papers are reviewed under the supervision of the Editors with the assistance of the Series Advisory Board and are selected to maintain the integrity of the symposia; however, verbatim reproductions of previously published papers are not accepted. Both reviews and reports of research are acceptable, because symposia may embrace both types of presentation.

Contents

Preface vii

CONSERVATION, DEGRADATION, AND CHARACTERIZATION OF PAPER

1. **Permanence and Alkaline–Neutral Papermaking** 2
   D. J. Priest

2. **Critical Evaluation of Mass Deacidification Processes**
   for Book Preservation 13
   David N.-S. Hon

3. **Graft Polymerization: A Means of Strengthening Paper and**
   Increasing the Life Expectancy of Cellulosic Archival
   Material 34
   C. E. Butler, C. A. Millington, and D. W. G. Clements

4. **Damaging Effects of Visible and Near-Ultraviolet Radiation**
   on Paper 54
   S. B. Lee, J. Bogaard, and R. L. Feller

5. **The Effect of Variations in Relative Humidity**
   on the Accelerated Aging of Paper 63
   Chandru J. Shahani, Frank H. Hengemihle,
   and Norman Weberg

6. **A Reexamination of Paper Yellowing and the Kubelka–Munk**
   Theory 81
   Harald Berndt

CONSERVATION AND DEGRADATION OF TEXTILES

7. **The Stabilization of Silk to Light and Heat:**
   Screening of Stabilizers 94
   M. A. Becker, S. P. Hersh, and P. A. Tucker

8. The Conservation of Silk with Parylene-C 108
   Eric F. Hansen and William S. Ginell

9. Historic Silk Flags from Harrisburg 134
   M. Ballard, R. J. Koestler, C. Blair, C. Santamaria,
   and N. Indictor

10. Long-Term Stability of Cellulosic Textiles: Effect of Alkaline
    Deacidifying Agents on Naturally Aged Cellulosic
    Textiles 143
    N. Kerr, T. Jennings, E. Méthé

11. Heat-Induced Aging of Linen 159
    Howard L. Needles and Kimberly Claudia J. Nowak

12. Treatment of Tapa Cloth with Special Reference to the Use
    of the Vacuum Suction Table 168
    Sara Wolf Green

CHARACTERIZATION AND PRESERVATION OF TEXTILES

13. Identification of Red Madder and Insect Dyes by Thin-Layer
    Chromatography 188
    Helmut Schweppe

14. Ultraviolet and Infrared Analyses of Artificially Aged
    Cellophane Film 220
    Laura DeSimone and Ira Block

15. Nondestructive Evaluation of Aging in Cotton Textiles
    by Fourier Transform Reflection–Absorption Infrared
    Spectroscopy 239
    Jeanette M. Cardamone

INDEXES

Author Index 254

Affiliation Index 254

Subject Index 254

Preface

EXTENSIVE RESEARCH HAS BEEN PUBLISHED on the chemistry and physics of paper and textiles. From the volume of available work, physical scientists must extract the information required by conservators to assist them in the preservation of fibrous materials. To this end, the Cellulose, Paper, and Textile Division of the American Chemical Society has sponsored four symposia since the mid-1970s on the preservation of paper and textiles of historic and artistic value. These conferences provided a forum where conservators and physical scientists could meet and discuss matters of mutual interest. Papers presented at the first three meetings have been published as chapters in three volumes of the Advances in Chemistry Series:

- *Preservation of Paper and Textiles of Historic and Artistic Value*; Williams, John C., Ed.; Advances in Chemistry 164; American Chemical Society: Washington, DC, 1977.

- *Preservation of Paper and Textiles of Historic and Artistic Value II*; Williams, John C., Ed.; Advances in Chemistry 193; American Chemical Society: Washington, DC, 1981.

- *Historic Textile and Paper Materials: Conservation and Characterization*; Needles, Howard L.; Zeronian, S. Haig, Eds.; Advances in Chemistry 212; American Chemical Society: Washington, DC, 1986.

This volume contains chapters from the fourth symposium.

The seriousness of problems related to the conservation of paper is already well recognized. In about 1850, paper became much more susceptible to deterioration because of the acidic nature of the products prepared by the manufacturing processes then being introduced. Today, steps are being taken to correct and prevent problems. The *Wall Street Journal* of March 6, 1989, reported that the publishing industry estimated that in 1990, 50% of all paper used in book publishing would be acid free compared with only 25% in 1989. According to a report in the March 13, 1989, *Chemical & Engineering News,* acid-free paper is estimated to last 300 years compared with an approximately 30-year lifetime for acidic paper.

The change to alkaline–neutral papermaking is laudable and will assist in the preservation of books published in the future. (The production of alkaline–neutral paper is surveyed in Chapter 1 of this volume.) However, the difficulty with books printed since 1850 remains. Methods of deacidifying paper are critically evaluated in Chapter 2, and the potential of graft copolymerization as a means of strengthening paper is described in Chapter 3.

Another problem conservators face is the deterioration of paper by exposure to light; it is discussed in Chapter 4. Paper is hydrophilic and may turn yellow over time. The importance of controlling the ambient atmosphere in which books are stored is brought out in Chapter 5, and yellowing is discussed in Chapter 6.

Unlike paper, textiles are made from a wide range of fibers formed from different types of polymers. Textiles are usually colored, and the type of dye used depends on the fiber. Thus, each fiber has its own unique set of problems. For example, synthetic fibers are less susceptible to insects than are natural fibers, whose potential for damage depends on the fiber and on the insect. Also, the rate at which fibers deteriorate when exposed to sunlight varies, depending on how they have been dyed and which type of dye has been used. Again, a fiber's susceptibility to a reagent depends on its organochemical nature. Different dyes are susceptible to different reagents as well. Thus, whereas some general rules can be applied to textile conservation, knowledge of the individual fiber, dye, and finish is important.

Currently, the vast majority of textiles being collected by museums are made from natural fibers, and attention is focused on these products in this volume. Silk is discussed in Chapters 7–9, and cellulosics in Chapters 10 and 11. Techniques that may be useful for the characterization of textiles to be preserved are described in Chapters 13–15.

One of the fascinations of studying textiles is that in addition to being manufactured from conventional fibers, they can be formed from other materials. Problems related to conservation of a particularly sensitive material, tapa cloth, are discussed in Chapter 12.

The authors wish to thank Sandy Brito for her valuable and prompt assistance with respect to the correspondence generated in the organization of the symposium and of this volume. We would also like to acknowledge the help we received from Cheryl Shanks and Donna Lucas of the ACS Books Department in the preparation of this book.

S. HAIG ZERONIAN
University of California
Davis, CA 95616

HOWARD L. NEEDLES
University of California
Davis, CA 95616

August 3, 1989

CONSERVATION, DEGRADATION, AND CHARACTERIZATION OF PAPER

Chapter 1

Permanence and Alkaline–Neutral Papermaking

D. J. Priest

**Department of Paper Science, University of Manchester Institute of Science
and Technology, Manchester, P.O. Box 88, Sackville Street, Manchester M60
1QD, United Kingdom**

The papermaking process is increasingly being modified so
that the sheet is formed in a neutral or alkaline aqueous
environment, rather than in an acidic one. Paper made in
this way is normally longer lasting because acid hydrolysis
of the cellulose can no longer occur. However, the reasons
for introducing the modified process are largely economic,
and the product may not necessarily meet specifications for
permanence and durability. This review describes the
technicalities of the economic advantages (including easier
fibre refining, increased filler content, the use of calcium
carbonate fillers, and the availability of cost-efficient
neutral sizes), the factors involved in making a change to
neutral/alkaline papermaking, and how all this impinges on
producing a satisfactory permanent paper.

Paper is essentially a bonded mat or felt of relatively small fibres to
which can be added, if required, fillers, wet strengtheners, coatings
and so on. Although a paper-like material can be produced from many
different polymeric fibres, paper itself is nearly always made using
fibres from natural sources, usually, but not exclusively of course,
from wood. These natural fibres all comprise polysaccharides of one
sort or another, predominantly cellulose, which are very hydrophilic
because they contain many accessible hydroxyl groups. The essential
adhesion between fibres is a consequence of hydrogen bonds formed
through these hydroxyl groups, as is the sensitivity of unmodified
paper to disintegration when wetted by water.
     In many of its uses, paper needs to have resistance to penetration
by aqueous fluids such as writing inks or the damping solutions used in
lithographic printing. The treatment given to the surfaces of fibres
to make them hydrophobic, which is usually done as the sheet is being
formed, pressed and dried on the papermaking machine, is known as
"internal sizing", to distinguish it from "surface size" applied on a
size press part way down the drying section of the machine.

0097–6156/89/0410–0002$06.00/0
© 1989 American Chemical Society

Since the early days of machine made paper in the first half of the nineteenth century, the most widely applied method of internal sizing has been the use of naturally occurring resinous materials ("rosins") in conjunction with an aluminium salt, usually aluminium sulphate (called "alum" by papermakers). Various forms of rosin sizes (rosin soaps, rosin emulsions, fortified rosins) have been developed over the years to improve the process, but these variants still involve the use of alum as a means of ensuring that fibres retain a layer of size.

Aluminium sulphate hydrolyses in aqueous solution to yield complex hydrated aluminium ions plus hydroxonium ions (1, 2), and hence a low pH. Papers made using alum/rosin sizing are often said to be "acidic", although this is rather imprecise terminology. A complete definition, following the related TAPPI standard method (3), is that paper acidity is the extent to which water-soluble materials in the paper alter the hydrogen-hydroxyl ion equilibrium of pure water causing an excess of hydrogen ions as measured by a commercial pH meter under specified conditions.

The important point is that the cellulose in these alum/rosin sized papers is susceptible to acid hydrolysis, which results in a lowering of the degree of polymerisation and, eventually, to a serious reduction in the strength of fibres and to complete embrittlement of the paper. Some recent work in the writer's laboratory suggests that when alum/rosin papers are made, the hydroxonium ions which lead to the degradation are adsorbed independently of aluminium ionic species (4).

In recent years, increasing attention is being paid by the paper industry to systems in which sizing is accomplished without the need to have the wet end of the machine running at acidic pH values. In these newer systems the pH may be around the neutral point, or be slightly alkaline due usually to the use of calcium carbonate filler (see below), so they are known as "neutral/alkaline". Papers made in this way do not yield acidic aqueous extracts and hence degrade more slowly (5, 6). Clearly, this is of great significance to those concerned with ensuring that important books and archival documents use paper expected to have a long life, and which will not lead in 30-150 years time to the enormous problems now being experienced in libraries and archives with paper made 30-150 years ago (7). However, it must be recognised that the reasons for introducing neutral/alkaline papermaking were not primarily associated with permanence; papers made in this way do not necessarily meet all the requirements for permanence and durability. Also, the alum/rosin acidic sizing method has been such a dominant force in papermaking that many other features of the process have been designed around it and adapted to it; the often used term "alum/rosin sizing system" is entirely appropriate. Making the change to neutral/alkaline papermaking nearly always involves, as we shall see, much more than throwing a switch or opening a valve.

In a previous publication in this series (8), Hagemeyer set alkaline papermaking in the context of future demand for paper, and dealt briefly with some of the technical consequences. Since then, more mills have converted to the new method, and the aim of this chapter is to inform the reader in some detail about the reasons for changing to neutral/alkaline papermaking, some of the consequences for the production and properties of paper, and how the change impinges on permanence and durability. It is important for those concerned with conservation and permanence to be able to communicate with papermakers and others with an awareness of relevant problems. Where possible, literature is cited, but a complete review is not intended, and some of the comments arise from the writer's past involvement in some of the industrial aspects of neutral/alkaline papermaking.

REASONS FOR CHANGING TO NEUTRAL/ALKALINE PAPERMAKING. As in most industrial change, the chief incentive is economic, and we need to look at ways in which the neutral/alkaline process gives rise to savings in the cost of production. Four main areas are involved: the fibre furnish, mineral fillers, the sizing system and the papermaking process itself. Although for convenience these will be discussed in turn, it should be noted at the outset that there is a great deal of interaction between the various aspects.

FIBRE FURNISH. It is well established (9) that when fibres are beaten or refined at a neutral or slightly alkaline pH the efficiency of the process is greater than at the acidic pH of around 4.5 common in alum/rosin systems. (When running an alum/rosin system it is inevitable that much of the stock preparation part of the mill operates at low pH because most of the water used is recycled from the wet-end of the paper machine).

The increase in refining efficiency means, for example, that a given level of strength in the paper can be obtained for a lower expenditure of energy. This is a major fundamental economic incentive for converting to neutral/alkaline papermaking, because large amounts of expensive energy are consued in refining fibres (10). This basic advantage can be exploited in different ways, depending on the particular product being made and market requirements (9). For example:

a)   The composition of the fibre furnish can be altered. The proportion of hardwood pulp might be increased, for instance, to give a product with the same strength as before, but with improved formation and opacity. Some cheap, relatively weak, bleached mechanical pulp might be introduced, or the proportion already used increased, again giving better uniformity and opacity, and a lower apparent density, but without loss of strength. This latter trend, of course, would not be acceptable in a permanent grade of paper.

b)   The potentially improved strength can be offset by increasing the amount of mineral filler in the paper, and this is a common route to follow, because fillers are usually much less expensive than the fibrous raw materials they replace, whilst at the same time properties such as brightness and opacity are improved. This important aspect is discussed more fully in the next section.

c)   A product of similar composition can be made but simply using less energy in refining.

In fact, these three approaches are not mutually exclusive, and a mill would need to consider how to combine changes to optimise financial savings whilst producing a paper acceptable in quality to the particular market being served.

FILLERS. In addition to being able to use more filler, a very important feature of running neutral/alkaline is the capability of greatly increasing the choice of mineral filler. This is because it becomes easily possible to use fillers constituted from calcium carbonate ($CaCO_3$), of which there are many different types. In alum/rosin systems, the pH is low enough for chemical reaction with the $CaCO_3$ to occur, producing troublesome evolution of CO_2 gas, causing froth and foaming and altering the ionic constitution and pH of the wet-end circuits.

$$CaCO_3 + 2H_3O^+ \longrightarrow Ca^{2+} + CO_2 + 3H_2O$$

Some attempts have been made in the past to overcome this difficulty by pre-treating the slurry of carbonate filler with special starches or water soluble polymers in order to protect the filler particles from acid attack for long enough to avoid foaming etc. (11); if the treated carbonate slurry is added at a suitable point, the dwell time in the acid environment is relatively short. Although these systems can work well if properly set up and controlled, they have not found wide application, largely being superseded by the advent of cost effective neutral sizes, which also avoids the cost of the protecting starch or polymers. However, a parallel development is the availability of rosin size emulsions which are effective at higher pH's (i.e. just on the acid side), and at least one mill in the UK has been taking this approach to using low additions of alum with carbonate filler (12).

Once again, the advantage of being able to use carbonate fillers can be realised in many different ways, depending both on the product and market requirements, and also on the availability and cost of filler supplies. Calcium carbonate fillers are produced either by controlled comminution of naturally occurring materials differing as widely as chalk, limestone or even marble, or by a chemical process leading to "Precipitated Calcium Carbonates", or PCC's. Within each type there are a range of products, varying in particle size and distribution, particle shape, and brightness. Different materials are produced at different locations throughout the world, so affecting detailed local economics. In Europe, there is a plentiful supply of inexpensive ground chalk filler, and there is usually an incentive to replace some or all of the clay (used in an acid sizing system) with chalk, and to increase the total filler content. However, due regard must be paid to relevant properties of the paper; e.g. large proportions of chalk filler will increase the oil absorptivity of the paper and hence its behaviour in printing processes. Also, although the more efficient alkaline beating will generally allow retention of strength at higher filler levels, the relative values of different types of strength can change, leading to possible difficulties in use. For example, if burst and tensile strength remain unaltered, but the paper is not as stiff as before, there is a danger that sheets will not feed properly into printing machines.

In the USA, where there is not the same supply of cheap ground chalks, it may be cost effective to use the more expensive precipitated carbonate, especially if it can be prepared in the mill, as is often the case. Through proper control, it is possible to make fine particle sized uniform products of high brightness, giving the possibility of replacement, at least in part, of very expensive speciality filters such as titanium dioxide (13). Use of PCC can sometimes also be justified, even where supplies of cheap chalk are available, when making products involving the use of TiO_2.

In complete contrast, the choice might be a lower brightness coarser filler where the main aim is cheapening a product without much affecting its optical properties, i.e. employing carbonate only as a filler.

Using appropriate techniques, and for suitable products, it is now possible to make satisfactory papers containing 25-30% w/w of chalk filler, although 15-20% in general neutral/alkaline printing and writing grades is probably more common. Such high levels of filler are not needed for supplying an "alkaline reserve" in permanent grades of paper; the American National Standard for permanence of paper for printed library materials proposes a minimum of 2% as calcium carbonate (14). Although the presence of excess filler is unlikely to be detrimental to permanence, it could mean that the mechanical properties of the paper do not meet the requirements for initial durability - such as those specified in the standard.

SIZING. Clearly, the key to the increased use of neutral/alkaline systems is the availability of suitable cost-efficient sizes. This has come about through the development of synthetic materials which are designed to form chemical covalent bonds with the hydroxyl groups in the surfaces of fibres (13, 15). In addition to the reactant group, the size molecule also has a hydrophobic portion, usually consisting of short alkyl chains. The two types of size in most common use are alkyl ketene dimers (AKD) or alkyl succinic anhydrides (ASA); Figure 1 shows the intended sizing reactions.

In practice several problems have had to be overcome before this apparently attractive method of sizing could be implemented efficiently. Since the ketene or the anhydride have to react with hydroxyl groups, they will also react readily with water; i.e. the molecules are hydrolysed to give non-reacting carboxylic acids (Figure 2). Some means must therefore be found to permit addition of the sizes to the wet-end of a paper machine, and then to ensure that they are retained within the wet paper web in such a way that an adequate size film is deposited on fibres in the dried sheet. This is made more awkward by the essentially hydrophobic nature of the molecules. The means adopted is to prepare emulsions of the sizes, often using cationic starch as a stabiliser and retention aid.

The storage stability of these reactive synthetic size emulsions is also of practical importance; AKD sizes tend to be delivered by the manufacturer in emulsion form, whilst ASA is emulsified on site shortly before pumping it into the wet-end. This is an area where much confidential manufacturer's expertise comes into play.

At one time, difficulties were encountered with ensuring that the desired degree of sizing developed in a reasonable time, especially with AKD's. With rosin/alum, sizing is complete in the reel at the end of the paper machine, but with some early AKD sizes, water resistance continued to develop for some days after the paper was made, making quality control difficult if not impossible. With newer grades of AKD this problem no longer arises, provided care is taken to ensure that temperatures in the drying section of the paper machine are high enough

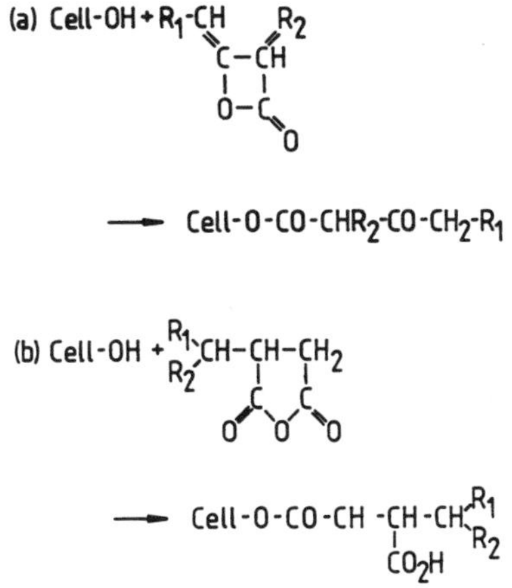

(a) Cell-OH + R$_1$-CH, R$_2$, C-CH, O-C, O

$\longrightarrow$ Cell-O-CO-CHR$_2$-CO-CH$_2$-R$_1$

(b) Cell-OH + R$_1$/R$_2$CH-CH-CH$_2$, C, C, O, O, O

$\longrightarrow$ Cell-O-CO-CH -CH-CH, R$_1$/R$_2$, CO$_2$H

Figure 1. Postulated sizing reactions of (a) alkyl ketene dimers and (b) alkyl succinic anhydrides. R$_1$ and R$_2$ are short alkyl chains. Cell-OH represents cellulose.

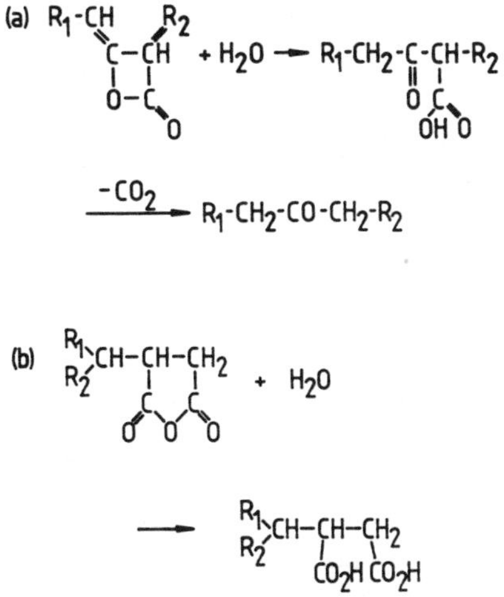

(a) R$_1$-CH, R$_2$, C-CH + H$_2$O $\longrightarrow$ R$_1$-CH$_2$-C-CH-R$_2$, O-C, O, C, OH O

$\xrightarrow{-CO_2}$ R$_1$-CH$_2$-CO-CH$_2$-R$_2$

(b) R$_1$/R$_2$CH-CH-CH$_2$, C, C, O, O, O + H$_2$O

$\longrightarrow$ R$_1$/R$_2$CH-CH-CH$_2$, CO$_2$H CO$_2$H

Figure 2. Hydrolysis (non-sizing) reactions of (a) alkyl ketene dimers and (b) alkyl succinic anhydrides.

in the correct positions. The actual mechanism of sizing with these new materials has been the subject of recent studies (16, 17), it seems that only a small proportion of the size actually reacts chemically with the cellulose fibre surface, but that it is essential to add the excess initially. The synthetic sizes tend to be more expensive than rosin, so, neglecting other economies, in a straight replacement they need to be used in smaller quantities; typical addition rates might be 0.5 - 1% w/w of dry fibre. The amount consumed will be related to the type of pulps being used, to the degree of beating and refining, and to the amount and type of filler.

No specific information on the possible effect of the synthetic sizes on permanence is available, and they are not mentioned in the standard (14), but it seems unlikely that they would be deleterious. Presumably they have been used in commercial grades of neutral/alkaline paper subjected to accelerated ageing tests. In terms of general effects on paper, the synthetic sizes have a tendency to reduce the surface frictional properties of paper to a greater extent than rosin. For example, this makes it more difficult to stack piles of cut sheets without slippage. However, the effect is less noticeable when high proportions of chalk filler are used, because the 'blocky' particles increase friction.

THE PAPER MAKING PROCESS

The repercussions of running neutral/alkaline on the total process are widespread and merit a separate article. Whilst in the present context it is not necessary to deal with the topic in detail, to understand the relationship of permanent paper to neutral/alkaline papermaking as a whole, the reader needs to have some appreciation of what it means to change to the new process. In particular, this understanding is very useful for effective communication between manufacturer and user.

To meet this need, brief information is given on each of the relevant main areas of the process:
- the stock preparation, approach flow and wet-end systems.
- formation and drainage on the wire section.
- wet pressing.
- drying.
- size press treatment.
- re-use of broke and waste paper.
- effluent treatment.

STOCK APPROACH FLOW AND WET END SYSTEMS. The papermaking stock being pumped to the headbox contains a number of additives extra to the fibre, filler and size. According to the grade being made, these may include retention aids, dyes or pigments, optical brighteners, pitch control agents and wet or dry strengthening aids. Switching to neutral/alkaline is likely to affect the performance of any or all of these additives. For example, the hue and depth of colour given by dyestuffs is often related to pH, so running alkaline is likely to require changes in the type of dyestuff being used (18). In fact, in planning to change from acid to neutral/alkaline conditions any mill will need to review the nature of all its ancillary materials, and how they are likely to function under the new circumstances. Increased filler levels make choice of retention aids particularly important (19).

Means of ensuring the cleanliness of wet end circuits and approach flow systems will also need attention. Warm suspensions of cellulose fibres and starch are excellent breeding grounds for organisms producing various sorts of unwanted slimes and deposits, and it is customary to add suitable inhibitors. Such growth is often sensitive to pH, and when pH changes different strains of bacteria and fungi become active, requiring different types of slimicide.

To reduce the consumption of fresh water and to minimise volumes of effluent needing to be treated, mills generally seek to run with systems which are as nearly 'closed' as is practicable; i.e. as much water as possible is continuously recycled within the mill. Consequently, increasing quantities of soluble substances are retained within the wet end system. A pH change may once again affect the nature and quantity of these materials, although in general running alkaline may be beneficial here.

Changing to neutral/alkaline conditions can also result in fewer corrosion problems in the wet end systems, a further aspect which has to be taken into account (20).

DRAINING ON THE WIRE SECTION. The differences obtained when beating or refining under neutral/alkaline conditions can produce some unexpected changes in the behaviour of the stock as it drains on the forming fabric of the paper machine - a specially critical region of the process in terms of product quality and uniformity.

At increased levels of carbonate filler, which tends to be more hydrophobic than clay, water drains more readily from the stock, and this is generally an advantage which allows various useful changes. For example, the consistency in the head box can be reduced, to give improved formation and uniformity in the sheet without having to slow down the machine to cope with the extra volume of water needing to be drained.

A key visual indicator of drainage behaviour is the so-called 'dry-line', which is the position down the wire where the sheet of draining stock loses its wet gloss and becomes matt in appearance. When running alkaline the wet web can remain glossy farther downstream, even though the actual solids content has not altered (21). Once this is appreciated, there should be no problem, but it is yet another example of unexpected change.

WET PRESSING. Where wet presses have a plain roll in direct contact with the wet paper web, serious difficulties have been encountered, attributed to hydrolysed and poorly retained neutral size residues (22). These are deposited on the surface of the roll, building up a film to which the wet web adheres, causing wrap rounds and web breaks. If no solution can be found, this would be a big enough problem to preclude running neutral/alkaline; particular attention needs to be paid to minimising pre-hydrolysis and maximising size retention.

DRYING. This is an area where neutral/alkaline papermaking can show advantages, especially when using increased quantities of carbonate filler (21). The hydrophobic nature of the filler, combined with the reduced proportion of hydrated fibre in the web, both mean that drying is more energy efficient. Reducing steam consumption or increasing machine speeds (where maximum speed was previously dryer-limited) both represent substantial potential economies.

SIZE PRESS. Grades of printing paper, including those for use where permanence and durability are important, are usually treated with starch, or some other water soluble polymer, at a size press part way down the drying section. The degree of internal sizing present in the pre-dried sheet entering the size press helps to control the amount and extent of penetration of surface size picked up. It is very likely when running neutral alkaline that the absorption characteristics of the web at the size press will be different, either because sizing has not developed to the same extent or because of a higher filler level (23) so the process is likely to need modification here too.

RE-USE OF BROKE AND WASTE PAPER. This is an extensive and complex area which can only be touched on here. (Broke is the term used for paper made on a machine which does not end up in the finished reel; edge trims, waste at reel changes etc.). Examples of relevant factors are:
a)   It is very difficult to run acidic broke or waste in an alkaline system (21).
b)   One incentive for changing from acid to neutral/alkaline conditions can simply be the need to run the paper or board machine using a proportion of broke from, for instance, an associated paper coating plant employing a calcium carbonate pigment.
c)   Papers sized with neutral reactive sizes are sometimes more difficult to disintegrate in broke pulpers.
d)   In a multi-machine mill where only one machine was making an archival grade, it would probably be necessary to segregate the broke from that machine, to ensure that broke from other machines was not used. This is often not a normal procedure and generates added cost.

EFFLUENT TREATMENT. Again, this is a complex matter, and circumstances will differ from mill to mill, depending on product and situation. Two features of some importance are:
a)   Running neutral/alkaline is said to allow mills to operate with a highly closed system because there is less dissolved material to accumulate, and this means a lower volume of effluent to treat.
b)   Absence of alum, which is a very effective flocculant for suspended solids, means that an alternative cost-effective synthetic flocculant will be required for the effluent treatment plant.

PROPERTIES AND PRICE. Because of the wide range of grades of paper and paperboard being made both acid and neutral/alkaline, it is not possible to generalise on the effect of a process change on paper properties. If a single grade is made by either method, and filler levels or furnish have not been changed, there are unlikely to be any distinctive differences in general physical properties (but recall the surface frictional effects mentioned above) - indeed, the new neutral/alkaline grade would be required to meet the same specification.

However, it is clear that in one very important respect the two papers would be different, and that is in their response to accelerated ageing tests. Where permanence is required, the advent of neutral/alkaline sizing has enabled satisfactory grades to be made in a way not previously possible. Running neutral/alkaline, as we have seen, is only one aspect of permanence; attention must still be paid to having the right quality of fibre (excluding all lignin for example), and including amounts of carbonate filler which will act as an alkaline reserve, but will not be present in sufficient quantity to adversely affect the properties of the paper (14).

Again, in considering the price to be paid for paper, it is vital that the correct comparisons are made. If a standard commodity grade of neutral/alkaline sized paper, made in large tonnages, meets the required specification for permanence, it should clearly cost no more than the corresponding acid sized paper. Even if it does not meet a set permanence specification, it is very likely to be considerably better in this respect than the acid sized grade it has replaced.

On the other hand, if a special archival grade is being made to order and a tight specification, in small tonnages, with a non-standard expensive fibre furnish, on a small slow machine which needs a special cleaning before the making, then the paper is likely to be equally expensive whether it is made acid or alkaline. Either way, it will still cost more than the high tonnage standard commodity grade.

CONCLUDING REMARKS

The prospects for improving the longevity of paper in books, documents and archives has been greatly enhanced by the introduction of practicable cost effective systems of neutral/alkaline papermaking. In many ways, the problems are now in the techno-economic and marketing fields - every effort needs to be made to ensure that those responsible for specifying the paper to be used for books, archives and so on, are fully informed about the merits of the various grades of neutral/alkaline paper now available for selection. The contents of this article are intended to make a positive contribution as a source of relevant information, because it is undoubtedly important for those concerned with purchasing paper for permanence to be aware of how such paper relates to the wide field of general neutral/alkaline papermaking.

LITERATURE CITED

1. Arnson, T.R.; Stratton, R.A. TAPPI (1983) 66, 72.
2. Strazdins, D. TAPPI (1986) 69, 111.
3. TAPPI Official Test Method, T 509 om-83 (1983).
4. Priest, D.J.; Farrar, M. 'Symposium 88' proceedings, Canadian Conservation Institute, 1988.
5. Tosh, C. Paper (1981) 195, 26.
6. Selawy, A.C.; Williams, J.C. TAPPI (1981) 64, 49.
7. Fifield, R. New Scientist (9 April 1987) 31.
8. Hagemeyer, R.W. In Preservation of Paper and Textiles of Historic and Artistic Value II; Williams, J.C., Ed.; American Chemical Society, Washington, 1981; 241.
9. Breslin, J. Paper Trade Journal 1985, 169, 57.

10. Smook, G.A. 'Handbook for Pulp and Paper Technologists', TAPPI/CPPA; 1982; 181ff.

11. Brooks, K.; Shiel, L.E.; Smith, D.E. U.S. Patent 4 272 277, 1981.

12. Roberts, F.J.; Wilson, C.M.W. In Trends and Developments in Papermaking; Evans, J.C.W., Ed.;   Miller Freeman Inc.: San Francisco, 1985; p 15.

13. Brink, H.G.; Gaspar, L.A. TAPPI Seminar Notes, Alkaline Papermaking (1983) 15.

14. Permanence of Paper for Printed Library Materials, American National Standard, ANSI 239, 48-1984, .

15. Alberts, A.M. TAPPI Seminar Notes, Alkaline Papermaking (1983) 35.

16. Davison, R.W.; Hirwe, A.S. TAPPI Alkaline Papermaking Seminar Notes (1985) 7.

17. Roberts, J.C.; Garner, D.N.; Akpabio, U.D. In Papermaking Raw Materials; Punton, V., Ed.; Mechanical Engineering Publications Ltd., London, 1985; Vol. 2, p 815.

18. Westmoreland, J.; Majunclar, A. In Trends and Developments in Papermaking; Evans, J.C.W., Ed.;   Miller Freeman Inc., San Francisco, 1985; p 99.

19. Atkinson, J.G. Paper Trade Journal 1982, 166, 30.

20. Bowers, D.F.; TAPPI (1986) 69, 62.

21. Beach, C.M.; TAPPI Seminar Notes, Alkaline Papermaking (1983) 75.

22. McNamee, J.P.; ibid, p. 73.

23. Bryson, H.R. Paper Trade Journal 1978, 162, 26.

RECEIVED February 22, 1989

Chapter 2

Critical Evaluation of Mass Deacidification Processes for Book Preservation

David N.-S. Hon

Wood Chemistry Laboratory, Department of Forestry, Clemson University, Clemson, SC 29634–1003

Four mass deacidification processes for book preservation; namely, the Library of Congress diethyl zinc process, Wei T'o nonaqueous process, Kopper's "Book Keeper" process, and Langwell interleaf vapor phase process, are critically evaluated, based on their chemical characteristics and effectiveness on deacidification.

The collections of America's libraries are deteriorating rapidly. Thousands of books have already disappeared; millions are in grave danger. On March 29, 1987, The New York Times published an article with an attention getting title: "Millions of Books are Turned to Dust - Can They Be Saved?" (1) This title summarized explicitly the actual situation in libraries and archives in North America and Europe. The inherent acidity of book papers is the major cause of deterioration. In order to salvage a book, acid must be neutralized with alkaline chemicals. The neutralization process is called deacidification. Ideally, the process will also deposit an alkaline reserve to prevent future acid attack. Many deacidification processes have been developed since the 1960s (2-4). Several of these have also successfully demonstrated their potential and the possibility of operating them on a pilot or a commercial scale. In this review, a critical evaluation of these mass deacidification processes is made. The problems of acid deterioration of modern papers, development of deacidification processes, their chemical characteristics and effectiveness, as well as the need for the development of an integrated paper preservation program are discussed.

The Fate of Modern Papers

The cave paintings of Paleolithic man, the hieroglyphics chiseled into the crumbling antiquities of ancient Egypt, and the rune-covered artifacts of Scandinavia and Northern Europe were to preserve forever their activities and cultural heritage. Today, for the same purpose,

0097–6156/89/0410–0013$06.25/0
© 1989 American Chemical Society

we record our knowledge, technology, activities and culture on paper.
Although the cave paintings of Cro-Magnon man which were made 35,000
years ago are still in good condition, ironically, modern paper which
was made less than one hundred years ago is crumbling in libraries
throughout the world.

The facts are that of the 20 million books and pamphlets in the
collection of the Library of Congress, as many as 30% are in such a
critical stage of deterioration that they can not be circulated
(5,6). A recent survey of the New York Public Library revealed that
nearly 50% of its more than five million books are on the brink of
disintegration (7). This phenomenon can be observed in any major
university or research library. Millicent Abell of Yale University
Library has estimated that as many as 76 million books nationwide may
literally be crumbling into dust, with more joining the list every
year (8). A study conducted by William Barrow sadly indicated that
97% of all books published between 1900 and 1949 would have a useful
life of fifty years or less (9).

Why are the cave paintings of thousands of years old in better
condition than modern papers? Why do many early books printed in the
15th century show no signs of serious deterioration? How can we
preserve our accumulated knowledge? Are the current technologies on
book preservation effective? Are the current practice on book
preservation acceptable?

The objective of this report is to attempt to answer these
questions. Evaluation of current technologies on mass deacidification
processes are the main thrust of this work. In addition, the need of
an integrated, complete book conservation program is discussed.

Before embarking on any discussion on mass deacidification, the
history and development of paper making and causes that lead to
deterioration of modern paper are reviewed.

History and Development of Paper Making (10)

Paper, one of man's most essential commodities, was first made in the
Orient about 2,000 years ago. Credit for the invention of paper has
been given to T'sai Lun, a member of the Imperial Guard and Privy
Councillor, who conceived the idea of making paper from old rags,
flax, hemp, rice stalks and tree bark (11). The Chinese macerated
fibers from these materials in water and drained the suspension on a
mold covered with silk cloth. The fiber mats were removed and dried
in the sun to form paper. This uniqueness is attested to by its slow
communication to other parts of the world: five hundred years to
reach Korea and Japan; six hundred years to Samarkand and the Arab
world; and one thousand years to Europe, and even later to America in
1690. During that period, rags of cotton, flax, jute, and hemp
comprised the sole source of raw materials used in paper manufacture.

From the advent of papermaking, the use of rags or bast fibers
grew rapidly prior to 1800, creating a shortage of papermaking raw
materials. It has been reported that during that period, even linen
shrouds from exhumed corpses were sold for papermaking (12). In
1719, Rene de Reaumur, a brilliant French scientist, suggested that
paper could be made from the fibers of plants without using rags or
linen. However, it took until 1764 for a German clergyman, Dr. Jacob
Scaffer, to make paper experimentally from a wide variety of plant

materials and to demonstrate that these vegetable fibers could be a substitute for rags, yet no interest was aroused apparently at that time. At the beginning of the nineteenth century, with the use of paper and printing presses increasing rapidly, the demand for paper outstripped the production of handmade paper, and mass production techniques were called for. In 1840, Scaffer's idea was picked up by a German bookbinder named Christian Volter. He developed a wood grinder to produce groundwood pulp in 1844 and patented it in 1847. This was the beginning of mechanical pulp production. This development rapidly increased the production of newsprint, although the pulp was poor in quality, especially in strength and durability, being inferior to present-day mechanical pulp. In 1851, two Englishmen, Hugh Burgess and Charles Watt, produced pulp from willow shavings boiled in a solution of lye, making the first soda pulp from wood. This was the beginning of chemical pulp production. The advantages of chemical wood pulp over mechanical pulps were soon appreciated. Following this invention, the sulfite process was invented by an American chemist, Benjamin Tilgman, in 1867. The sulfate (kraft) process was invented in 1889 by a German chemist, Carl Dahl of Danzig. From this we can realize that within only a few years, a revolutionary change had taken place in the pulp and paper world.

Now, about one hundred years has elapsed. The fundamental pulping and papermaking principles established then still remain the basics of today's modern papermaking. Advances in engineering and technology have made it possible to produce increasingly larger tonnages and a vast variety of paper products by very cost-effective methods. Today, world consumption of paper amounts to about 200 million tons annually, of which 94%-95% is produced from wood, and the remainder mainly from other vegetable fibers. Ironically, the historical event seems to be repeated today; under the severe constraints of pollution control and energy conservation, we are experiencing a shortage of fiber raw materials. The result of this is the production of high-yield fibers that retain high amounts of lignin (or even extractives) in the pulp fibers. And the newest development in this area is the appearance of thermomechanical pulps that obtain pulp yields as high as 100%.

With the development of the chemical pulping technology of wood at the end of the nineteen century, animal gelatin, which had been used as the sizing agent, was replaced by alum. The main advantages of using alum-rosin sizing are that it does not need to be purified; when it is mixed into a slurry it acts to distribute rag or wood fibers evenly in water, and thus contributes to the making of an even textured paper; and paper sized in it dries quickly (13). The shortcoming of alum-rosin sizing is that slurries to which it is added, regardless of their makeup, yield acidic paper (14,15). For more than a century nearly all book papers have been sized this way.

Ironically, modern papers, which were invented after the revolutionary changes of one hundred years ago, have serious problems--either in permanence or in durability. Most paper produced from wood fibers does not show the permanence of rag papers and discoloration is more critical than in rag papers. It has been noted that modern writing papers and books tend to have a much shorter life expectancy than those manufactured one hundred years ago.

Permanence of Paper (16-18)

The permanence of paper is determined by "internal" and "external" factors (19). The internal factors are established during manufacture of paper and include kind and quality of the fiber, sizing materials, coatings and presence of acidic and metallic compounds, and other components of the sheets. The external factors are related to conditions during storage or use, e.g., temperature, relative humidity (which determines the moisture content of the paper), light, and contaminants in the atmosphere. The manufacturer is responsible for production of a paper having potential permanence; the user is responsible for storage of the product under conditions which are requisite for long life.

A lack of permanence in paper can be manifested by discoloration, loss in strength and change in chemical properties, such as increase in copper number, decrease in alpha cellulose, and decrease in degree of polymerization (DP) of the cellulose. In evaluating permanence, consideration should be given to changes in both the cellulose, hemicelluloses and lignin (chemical changes) and paper (physical properties).

Although many investigators have been concerned with permanence, the factors influencing the stability and permanence of paper have not been fully established. Both the quality of the cellulose fibers and the nature of the nonfibrous components contribute to the character of the finished sheet. Owing to the complex composition of papers and variations from grade to grade, many types of degradation are possible. For the cellulose fibers, these include hydrolysis, oxidation, cross-linking, change in crystallinity, photolysis, photo-oxidation, thermal decomposition and even microbiological attack under certain conditions (20). These factors combined with initial paper quality will severely reduce the permanence of paper.

Acid Formation

One of the most significant factors that contribute to deterioration and embrittlement is the introduction of acid elements into the paper (21). Acid catalyzed hydrolysis of cellulose causes 80-90% of paper deterioration (22). The rate of hydrolysis depends on various factors, including the nature of the cellulose and the conditions under which the paper is stored. Under most conditions, however, most papers deteriorate progressively. The aging characteristics of acidic paper are discoloration, embrittlement and steadily increasing fragility. Lignin-rich high yield papers suffer particularly badly.

The acid problem in libraries is much more acute than it was half a century ago. Books of the 1920's and 1930's are now entering a critical period. For the most part they can still be restored, but if they are not also protected from acid attack they will nevertheless become unusable in the foreseeable future (23).

The chief source of acidity in paper is the alum used in the papermaking process. Paper (largely cellulose) is a very hydrophilic substance, and its surface has a high specific energy. Thus, water readily wets paper surfaces. The very porous structure of paper makes it act like a sponge in the presence of liquids. Hence, various chemicals have been developed to make paper reasonably water repellent, so that paper can be printed or written with inks. The process in which a chemical additive provides paper with resistance to wetting and penetration is called sizing (13). Papermaker's alum $[Al_2(SO_4)_3 \cdot 14H_2O]$ is the most popular chemical used in sizing (14,15). It is introduced into the furnish to precipitate rosin size; to retain resins, starch, or pigments; and to control pitch (24). Unfortunately, aluminium sulfate is sensitive to humidity and undergoes hydrolysis to generate sulfuric acid (25). The sulfuric acid generated decomposes the cellulose molecule by breaking glycosidic bonds. The chain-shortening of cellulose leads to a corresponding decline of the mechanical strength of the paper.

Additional sources of acid contamination include carboxyl group in the cellulose, acidic carbohydrate gums, bleach and pulping chemical residues, and some constituents of coating colors. Acids may be introduced after the paper is manufactured by atmospheric contamination, particularly sulfur dioxide which is absorbed by the paper and, in the presence of moisture, will generate sulfuric acid.

Development and Need of Book Deacidification

If acid is present in paper, hydrolysis of cellulose fibers would be inevitable. Hence the removal of acid from paper is imperative. The first attempt to stabilize paper by a deacidification treatment was undertaken by Sir Arthur Church in 1891 using a solution of barium hydroxide in methanol to deacidify the backing of Raphael cartoons. In the early forties, Barrow recognized that the built-in acidity of paper had a greater effect on its deterioration than atmospheric contamination and advocated the deacidification of documents (26). Since then, a number of workers throughout the world have contributed to much of our present knowledge of deterioration due to acidity and of deacidification processes (2-4). It is known today that the inherent acidity is responsible for eighty-five to ninety percent of destruction in book papers, causing them to lose an estimated fifty percent of their fold strength every 7.5 years and ultimately reduces them into a state of embrittlement in which they will break at a touch (27). It has been observed by Smith (28) that the pH value required for stable, permanent paper has altered with the passage of time. In the early twentieth century, permanent book papers required a pH of 4 (hot water extract). In 1928, this figure was set at 4.7 or preferably higher. By 1935, it was realized that low pH was a cause of early deterioration in paper and the figure was raised to a minimum pH of 5 for good quality book paper. In 1937 Grant observed that for permanent paper the value of the hot water extract should not be less than pH 6. Lewis, in 1959, stated, on the basis of tests, that papers in good condition had a pH of 6.3 and 6.5. Barrow later also stated that the most stable papers show a pH of about 7 (cold extraction) while the least stable show a pH of 5, and that a pH of about 7 is desirable for maximum conservation. Smith also suggests that the most desirable pH value is 7, i.e., neutral. An evaluation by Kathpalia of papers dating from the fourteenth to nineteenth centuries has shown that papers with pH above 6.7 are in excellent condition while those with pH ranging from 6.2 to 6.7 are in good condition, and that all these papers are free from fungus stains.

As a result of the above studies, a number of processes have been developed for introducing a strong alkaline base to react with the acid in paper to form neutral salts. The process also deposits a neutral substance or buffer on the treated product that will resist future acid contamination from any source. The various processes developed may be classified as aqueous deacidification, nonaqueous deacidification and gaseous deacidification (5,29). It should be borne in mind that deacidification is effective for treating books which are already in the collections that have not yet been dangerously damaged by acid. Brittle books are beyond saving by deacidification only. For such books, a paper strengthening treatment has to be conducted also.

Aqueous Deacidification

Aqueous deacidification technique consisted of immersing paper in, or spraying it with, an aqueous solution of barium, calcium or strontium bicarbonates or hydroxides. The immersion time varied from five seconds to about two minutes. The wetted paper is then dried. As a result, carbonates of these metals are deposited on the treated paper. Pioneering work was done over fifty years ago by Schierholtz at the Ontario Research Foundation in Toronto. The process was patented in 1936. To increase the neutralizing potential of the chemicals used, Schierholtz recommended carbon-dioxide treatment to convert hydroxides to carbonates and that the suspension of carbonates be allowed to settle on the treated sheets. He reported that a more concentrated bicarbonate solution could be prepared by using carbon dioxide gas under pressure and advised that the pH of a water extract from paper treated by this process should exceed a value of 6.5 and that a deposit of up to 2 percent by weight might be required for stabilizing groundwood papers like newsprint. Based on the patent, Barrow of Richmond, Virginia, developed a two-step process (26), in which materials were first neutralized with calcium hydroxide and then alkalized with calcium bicarbonate. Many other approaches have been devised since that time, including the use of magnesium bicarbonate by Gear of the National Archives in Washington. His one-step process, developed in 1957, is the form of aqueous deacidification still practiced by many small libraries and archives. In 1978 this process was further improved by substituting the more soluble magnesium hydroxide for magnesium bicarbonate (30). The Library of Congress has used the aqueous deacidification method in the past for much of its paper and manuscript deacidification because of its safety and reliability (31).

Aqueous deacidification is an effective means to remove acid. Some of the acid and impurities can be washed out during the treatment. However, it is a painstaking process. Because water will swell paper, it will damage bindings. It requires taking books completely apart, soaking them in the solution, drying, and then rebinding. Moreover, water can be very damaging to certain inks used for color or for writing. Hence, the treatment require special handling skills for critically embrittled papers. For these reasons, a nonaqueous deacidification treatment has been developed to avoid such difficulties.

Nonaqueous Deacidification

A number of investigators have voiced the theory that a deacidification solution containing organic solvents might provide a remedy for difficulties caused by the use of water. Similar hopes were expressed by the International Institute for Conservation of Historic and Artistic Works in 1968, when its Committee for Paper Problems reported that "a nonaqueous means of deacidification that would not be harmful to paper, pigments and the various media must be developed." As a matter of fact, nonaqueous deacidification began in England about the end of the 19th century. Barium hydroxide in methanol was used as a deacidification agent in the Victoria and Albert Museum. Baynes-Cope of the British Museum reinvented this method in the middle of the 1960s (32).

Considerable research is being devoted to improve existing techniques, to accelerate the deacidification process, render it applicable for the treatment of bound volumes and reduce the cost of operations. Nonaqueous deacidification has been improved in the past ten years and has reached commercial success. The new process is conducted by introducing "liquified gas" into a chamber containing books. Once the acid ions in the paper have been neutralized, the "gas" is pumped out of the compression chamber and returned to a storage tank. The books can then be dried and return to the shelves. Since a nonaqueous deacidification process does not require books to be unbound as does an aqueous process, its development dramatically reduced the cost of operation.

Nonaqueous deacidification treatments involve the use of a nonaqueous solution containing a deacidification agent and an organic solvent. The advantage of using organic solvents is that they are available as liquids over a wide range of temperature and may be blended to obtain the desired working requirements and properties. One of the major features of this method is the rapidity of penetration of chemicals into paper and the extreme rapidity of drying the paper even at room temperature. This can avoid the problems of drying and crocking.

On the other hand, all the organic solvents used are either flammable or toxic or expensive. Some are poisonous, or hazardous to health, and others dissolve, or cause feathering in, dyes and inks used on paper. Substances such as magnesium acetate, barium hydroxide, cyclohexylamine and its carbonate and acetate and magnesium methoxide have been tested for their safety and effectiveness.

Magnesium methoxide has been found to be a very effective neutralizer. However, on a damp day or with a damp paper, the solution tends to precipitate prematurely and leave surface deposits on the treated paper. Methyl magnesium carbonate is also effective but much less sensitive to water. Both these products produce adequate alkaline reserves in paper. Since methanol is used as the solvent, the deacidification should be conducted in a well-ventilated hood.

The most successful nonaqueous deacidification treatment is the Wei T'o process which uses solutions of methoxy magnesium methyl carbonate in alcohol and Freon as the starting materials (33). The spray deacidification of the solution on books is now in use at the

Princeton University Libraries and at the British Library. The mass
deacidification of the Wei T'o process has been used in the National
Library and Public Archives of Canada (34). Just like any method,
Wei T'o process does have its limitations. Details of the chemical
process and its limitations will be discussed elsewhere in this
report. A similar process using magnesium alkoxides and alkoxy
carbonates dissolved in mixtures of methanol and Freons was developed
in the Bibliotheque Nationale at Chateau de Sable in France.

Gas Deacidification

In consideration of the damage of paper by liquids (solvents) as well
as the limitation and effectiveness of transfer chemicals into the
fibers, deacidification by gas has been devised. Although there are
a great many volatile organic compounds which appear to be sufficient
alkaline for vapor-phase deacidification, only a very few appear to
be practical. Kathpalia (35), at the Nehru Library in New Delhi,
exposed books to high concentrations of ammonia vapor for deacidification. However, it was found that ammonia is a too weak base to
completely neutralize strong acid in paper, and ammonia volatilized
from treated books in a few days. It has not been considered as a
permanent deacidification method.

Langwell (36-38) worked with cyclohexylamine carbonate as the
neutralizing amine salt. It has achieved some success. Although
this method is not strictly a mass deacidification process, it can be
easily applied to whole books, rather than single sheets. Detail of
its chemistry will be discussed in the subsequent mass deacidification section. The Barrow Laboratory investigated the use of several
amines and finally settled on morpholine as the most effective
vapor-phase treating agent (39). It appeared that the morpholine
penetrates books well and deacidifies them effectively. However,
this method has not been considered as a permanent method due to the
rapid drop of pH of treated products. The process does not introduce
a buffering chemical into books. They have to be redeacidified every
few years to maintain protection. Morpholine also may cause significant discoloration of papers. The Library of Congress has attempted
to use ethyleneimine as a gaseous treating materials, but no
significant improvement in aging characteristics was achieved, and
the paper was seriously discolored. Since then, the Library of
Congress had developed a promising vapor-phase process using an
organometallic compound, diethyl zinc, to deacidify books (40).
Detail of this process will be discussed in a subsequent section.

Chemical Characteristics of Mass Deacidification

With respect to the quantity of book materials to be neutralized in
libraries, it is too laborious to treat books in a single-sheet
operation. Only a mass scale deacidification process is acceptable.
Hence, several mass deacidification processes have been developed.
They are discussed further.

4. Chemicals used should be safe for personnel at the treatment
    site.
5. The treatment should not leave an odor in the books.
6. The treated books should be non-toxic to humans.
7. The treatment should have no deleterious changes in appearance,
    feel, or physical integrity of the books.
8. The treatment should not influence the properties of binding
    adhesives, leather and plasticized covers.
9. Chemicals discharged should not create environmental pollution
    problems.
10. Should achieve complete deacidification in a reasonable length
    of time.
11. Should uniformly neutralize and uniformly deposit alkaline
    reserve. The alkaline reserve should be two to three percent in
    concentration for maximum permanence.
12. The paper must be buffered at a pH of 7 to 8.5 for minimal
    effect on acid/alkali hydrolysis of paper and on inks and
    colors.
13. The expected life of the paper should be significantly improved.
    A minimum of a five time increase in accelerated aging lifetime
    must be achieved.
14. The treatment should be economically feasible for treating large
    quantity of books, i.e., > 50,000 volumes/year.

Evaluation of Different Mass Deacidification Processes

Although many mass deacidifications have been developed, it can be
stated that none of the technologies offered meets all the criteria.
Nevertheless, the DEZ process, Wei T'o process, Kopper Process and
VPD have demonstrated independently their potential for arresting
paper deterioration from acid. To what extent these processes meet
the criteria to be used for a practical mass deacidification process
is the question that has to be answered. Accordingly, a comparison
of the characteristic features of these processes is tabulated in
Table I. The evaluation is based on their neutralization chemistry
and effectiveness. The engineering design, safety and costs of these
processes are not considered in this evaluation.

DEZ Process. DEZ process has been developed and refined by chemists
at the Library of Congress since 1974. It is a very impressive
method of deacidifying book papers effectively and uniformly. There
is no doubt that the deacidification chemistry is workable. As shown
in Table I, the DEZ process is the process that met most of the
"ideal" criteria. In essence, the DEZ process uniformly and
consistently neutralizes all excess acid in the paper, leaves a
uniformly distributed alkaline reserve in all regions of the book
page and the paper fiber.
      Unfortunately, diethyl zinc is not a chemical with great
"stability" to work with. Because of its pyrophoric properties, DEZ
must be handled with great caution and subsequently cannot be
operated within libraries. This will limit potential library users.
      The end-product of deacidification is the conversion of sulfuric
acid in paper to zinc sulfate. Just like aluminum sulfate, zinc
sulfate is prone to dissociate into acid in the presence of water or

Table I. Comparison of Alternative Mass Deacidification Processes, As of January 1988

Criteria	Ideal	DEZ[a]	Wei T'o	Bookkeeper[b]	VPD[c]
Preselection of books	No	No	Yes	Minimal[d]	Yes
Predrying	None	Yes	Yes	None	None
Impregnation time	Short	Long	Short	Short	Very long
Treatment plant	Simple	Complex	Less complex	Simple[d]	Very simple
Effect on inks and colors	None	None	Some	Minimal[d]	Some
Effect on plastic covers	No	No	Yes	Minimal[d]	Yes
Neutralization	Complete	Complete	Needs verification[e]	Needs verification[e]	Partial
pH of treated paper	7.0-8.5	7.0-7.5	8.5-9.5[e]	8.0-9.0[e]	5.0-8.7
Alkaline reserve	About 2%	1.5-2.0%	0.7-0.8%[f]	2%[e]	None
Danger to health	None	Risk of fire	Uncertain[f]	Uncertain[f]	Uncertain
Impact on environment	None	Low	Uncertain[g]	Uncertain[g]	Low
Stage of development	----	Operating pilot plant (2 mo.)	Operating pilot plant (7 years)[h]	Lab tested pilot design	Commercial
Cost	----	Moderate to high[h]	Low to moderate	Low[h]	Low[h]

[a] Library of Congress' DEZ process.
[b] "Bookkeeper" submicron particle process.
[c] Langwell vapor phase deacidification process--distributed by Interleaf, Inc.
[d] Based on telephone conservation with Dr. J. J. Kozak of Koppers (Nov. 2, 1987). No independent assessment.
[e] No formal independent analyses have been made. Manufacturer's data indicates complete neutralization under laboratory conditions.
[f] Initial indications are good but no formal assessments have been made.
[g] Some concern about the future regulation of fluorocarbons used in these processes.
[h] Based on OTA analysis and extrapolation of limited cost data furnished by developer of each system.

moisture (55). This act is undesirable. The fate of zinc sulfate and the regeneration of acid in paper, unfortunately, has not been considered by the Library of Congress.

The reaction of diethyl zinc with water produces zinc oxide, and then zinc carbonate, as the alkaline reserve. These chemicals have antiseptic properties which may also prevent the growth of mold in paper. They may also improve the brightness of treated papers. However, it is also known that zine oxide is a photosensitizer (56) which may trigger photo-oxidation of treated papers to initiate a chemical chain reaction that will lead eventually to the formation of acidic products (57). Moreover, the interaction of zinc oxide and zinc carbonate with copper, iron and cobalt present in the paper and their subsequent effects on paper stability have not been studied.

Wei T'o Process. Wei T'o process is the only mass deacidification process currently available commercially. The "liquid gas" deacidi-fication solution of this process consists of methoxy magnesium methyl carbonate dissolved in a mixture of Freon and methyl alcohol. The process has been proven by six years of production operation at the National Library and Public Archives of Canada. It utilizes a less complex system than the DEZ process to achieve deacidification. National Library and Public Archives of Canada are satisfied with the deacidification results.

Nevertheless, as a mass deacidification process, it does have some limitations. The process requires a manual, item-by-item preselection procedure to remove books with plasticized covers, and books printed with unstable inks, to avoid color and ink transfer problems. Books with colored illustration (using unstable inks) can also cause problems. Testing has shown that some plastics used on modern paperback books react adversely. The finish might crack and flake. Examination of randomly selected books treated with the Wei T'o solution has shown nonuniform deacidification.

Although the deacidifying agent does not harm leather, leather bound books are not recommend for the treatment. It is anticipated that during pre-drying process, leather can be irreversibly damaged.

The other concern with the Wei T'o process is its use of Freon (chlorofluorocarbons). It is known that Freon vapor escaping to the atmosphere will lead to the depletion of the ozone layer in the stratosphere. Although the process is operated in a closed system, immediately after deacidification, books are stacked in shelves for drying. The escape of Freon to the atmosphere can be significant.

Wei T'o Associates announced that it has improved its nonaqueous deacidification solution to eliminate the ink instability problems, particularly in the graphic arts. The new formulation avoids the use of an alcohol co-solvent and the higher alkalinity of magnesium that may cause color changes or smudging on sensitive inks. Other solvents may replace chlorofluorocarbon solvents. This new formulation seems to solve all the problems that the Wei T'o process has. However, the refusal of Dr. Smith of Wei T'o to discuss or define its technology prevents us from evaluating its improved system. Hence, the improved system was not included in this evaluation, as Dr. Smith said, "If there is no definitive information about the idea or product, no consideration should be given it" (36).

Koppers "Book Keeping" Process. In view of the limitation of the Wei T'o process, chemists at the Koppers Company developed a "Book Keeper" process by dispersing submicron particles of basic metal oxides, hydroxides or salts of calcium, magnesium, or zinc, in a suitable gas such as Freon or liquid medium, so that the active chemicals can be transferred and deposited electrostatically on the surface of paper. It also does not require pre-drying of books as is required for both the DEZ and Wei T'o processes. The testing results appear satisfactory as shown in Table I. The major concern with this process is the distribution of the alkaline reserve on the paper. It appears the process deposits alkaline chemicals on the surface of paper and achieves surface deacidification. However, acid formed in the core of the paper is not neutralized. Koppers intends to prove the degrees of chemical penetration and neutralization of acid in the center layers by examination of the cross-section of paper by SEM. No reports have been released yet. The effects of the chemical process on paper, covers and inks have not been published. The Koppers process appears to be a unique one with great potential for mass deacidification with success. More testings are required. As with the Wei T'o process, the effect of Freon on the ozone layer of the stratosphere is a concern.

Langwell Vapor Phase Deacidification Process. Langwell's VPD process is a very simple process. Cyclohexylamine carbonate is inserted between book pages to achieve partial deacidification. However, this process does not create an alkaline reserve in treated products to prevent future acid attack. Cyclohexylamine carbonate will react with some plastic covers, inks and colors. Preselection of books is required. The odor and the carcinogenic nature of the vapor are also critical factors that limit the use of VPD process as a mass deacidification process (46-54).

To conclude, it is clearly evident that none of the mass deacidification processes just discussed meet the "ideal" criteria set forth in Table I. Without consideration of cost effectiveness and safety of operation, the Library of Congress' DEZ process seems to be the winner. However, as a mass deacidification process the Koppers "Book Keeping" process may have more potential due to its straightforward and danger-free operation.

Moreover, recently Book Preservation Associates and Lithco have announced new mass deacidification methods. These processes should be considered if data are available.

Integrated Complete Book Preservation Program

All of the deacidification techniques discussed above will remove acid and provide a deposition of alkaline reserve to protect further acid attack. Although these methods undoubtedly work, from least to most effective, they do not attempt to improve or restore the mechanical properties of the paper treated. Hence, mass deacidification is only a part of a complete mass preservation program that is needed by libraries and archives in order to extend the lifetime and usability of their holdings. Although deacidification of paper is a very important step in order to inhibit further deterioration, restrengthening of embrittled stocks is necessary as well as protection against detrimental oxidizing and biological attacks.

Accordingly, a complete book preservation program should consider (a) complete deacidification; (b) improving fold endurance, tear and tensile strength of paper; and (c) inhibiting oxidation.

The strengthening of paper can be done by deposition of polymers onto fibers or by grafting polymers onto fibers. Smith has mentioned briefly in many of his papers impregnating the paper in books with acrylic resin and ethyl hydroxyl-ethyl cellulose solution to increase fold endurance (no actual publications have been seen). Salz and Skrivanek have used polyvinyl butyrate for the same purpose. Sodium salt of carboxymethylcellulose (CMC) has been used by Raff and his co-workers (58). The Austrian National Library uses methylcellulose of low viscosity as the strengthening agent together with calcium hydroxide as the neutralizing agent (59-61). The British Library has grafted ethyl acrylate and methyl methacrylate onto book paper by gamma-irradiation (62). Vinyl monomers have been used at Clemson University with a heat- and light-induced graft copolymerization techniques to strengthening paper documents (63).

Antioxidants and photostabilizers may need to be added to treated books to avoid future acid generation in book papers.

Accordingly, without complete deacidifying and strengthening, a book conservation program should not be considered complete.

Recommendations

Based on the information available today on the DEZ process, Wei T'o process, Koppers process, and VPD process, it is very difficult to judge which process is the best suited to a mass deacidification process. In addition to the limitations of each process which has been discussed, data are lacking to support long-term effects on treated papers. In order to select a viable mass deacidification process, the following parameters have to be considered:

1. The effects of the reagents on the chemical properties of the paper components, such as cellulose, hemicelluloses, and lignin are not known. For example, the change in degree of polymerization, i.e., the chain length of cellulose polymer before and after treatments. The changes in functional groups, such as carbonyl, carboxyl, and conjugated double bonds in cellulose and lignin have not been considered, despite their importance on acid formation and discoloration reactions.
2. The optical properties of treated papers have to be studied. A colorimeter or reflectance spectrophotometer can be employed to examine the change in optical and color characteristics before and after treatments.
3. The mechanical properties have not been reported conclusively. Most of the available data provided was based on fold endurance testing. Although it is an important property that should be determined, studies of tensile, tear and burst strengths are important as well.
4. The oxidizability of treated papers due to heat, light and air pollutants need to be evaluated.

All these important parameters must be established for treated papers so that a meaningful evaluation can be made. Before making any decision on selecting an effective mass deacidification process, a comprehensive testing, which is based on the above-mentioned parameters, of the treated books from the DEZ process, Wei T'o process, Koppers process, and VPD process must be made by an independent group which is not associated with any of the developers of the deacidification process.

In addition to the deacidification chemistry, the final decision should also be based on the evaluation of the engineering design, safety of operation and cost factors which have not been considered in this paper.

Acknowledgment

This project was funded by the U.S. Congress, Office of Technology Assessment (OTC). This support is gratefully acknowledged.

Literature Cited

1. É. Stange, Million of Books Are Turning to Dust -- Can They be Saved? The New York Times, Book Review, March 29, 1987, p. 3.
2. G.M. Cunha, Mass Deacidification for Libraries, Library Technology Reports, 23, 363 (1987).
3. D. Baynes-Cope, Deacidification: A Statement from the Technical Committee, Journal of the Society of Archivists, 7, 402 (1984).
4. G.B. Kelly, Nonaqueous Deacidification: Treatment en Masse for the Small Workshop. Paper presented at the International Conference on the Conservation of Library and Archive Materials and the Graphic Arts, Cambridge, 1980.
5. J.C. Williams, Chemistry of the Deacidification of Paper, Bulletin of the American Group-IIC, 12, 16 (1971).
6. C.J. Shahani and W.K. Wilson, Preservation of Libraries and Archives, American Scientist, 75, 240 (1987).
7. R.E. Kingery, Permanent/Durable Book Paper, Virginia State Publication, No. 16 (1960).
8. Librarians Bemoan Costly Paper Chase, Greenville News, October 19, 1986, Section B.
9. W.J. Barrow, Deterioration of Book Stock - Causes and Remedies, The Virginia State Library, Richmond, Virginia, 1959.
10. D.N.-S. Hon, Yellowing of Modern Papers in Preservation of Paper and Textiles of Historic and Artistic Value II (J.C. Williams, ed.), American Chemical Society, Washington, D.C. 1981, Chapter 10.
11. Library of Congress, Papermaking: Art and Craft, Washington, D.C., 1968, p. 9.
12. E.W. Haylock, Paper: Its Making, Merchanting and Usage, The National Association of Paper Merchants, London, 1974, p. 8.
13. R.W. Davison and H.M. Spurlin, Sizing of Paper in Handbook of Pulp and Paper Technology, 2nd ed., Van Nostrand Reinhold Co., New York, 1970.
14. B.L. Browning, The Nature and Sources of Acid in Paper. Paper presented at the Institute of Paper Chemistry, Seminar for Conservators of Paper Objects, Appleton, Wisconsin, October 1971.
15. W.F. Reynolds and W.F. Linke, The Effect of Alum and pH on Sheet Acidity, Tappi 46, 410 (1963).

16. J. Byrne and J. Weiner, Permanence, The Institute of Paper Chemistry, Bibliographic Series No. 213, 1964; J. Weiner and V. Pollock, Suppl. I, 1970.

17. W.J. Barrow and R.C. Sproull, Permanence in Book Papers, Science, 129, 1075 (1959).

18. B.L. Browning, Analysis of Paper, Marcel Dekker, New York, 1977, Chapter 25.

19. B.L. Browning, Analysis of Paper, Marcel Dekker, New York, 1977, Chapter 8.

20. W.K. Wilson and E.J. Parks, An Analysis of the Aging of Paper: Possible Reactions and Their Effects on Measurable Properties, NTIS Report COM-74-11378, April, 1974.

21. J.S.M. Venter, The Aging and Preservation of Paper - A Development Study. Council for Scientific and Industrial Research, Pretoria, South Africa, September 1966, Chapter 1.

22. R. D. Smith, The History and use of magnesium alkoxides in the nonaqueous deacidification of books, documents and works of art on paper, Paper presented at the 1983 Annual Meeting of Internationale Arbeitsgemeinschaft der Archiv-, Bibliothek- und Graphikrestauratoren (IADA), The Hague, Netherlands, September 13, 1983.

23. Mass Deacidification at the NL, National Library News, 14, No. 3-4, 1982, National Library of Canada.

24. L.B. Miller, The Fundamental Chemistry of Alum and Its Application to Paper Manufacture. Tappi, 22, 141 (1939).

25. P.L. Hayden and A.J. Rubin, Systematic Investigation of the Hydrolysis and Precipitation of Aluminum (III) in Aqueous-Environmental Chemistry of Metal (A.J. Rubin, ed.), Ann Arbor Science, Ann Arbor, Michigan, 1974.

26. W.J. Barrow, Restoration Methods. A paper presented at the 6th annual meeting, Society of American Archivists, Richmond, Virginia, October 27, 1942.

27. J.M. Banks, Mass deacidification at the National Library of Canada. Paper presented at the annual meeting of the Society of American Archivists, Washington, D.C., September 1, 1984.

28. R.D. Smith, Paper Impermanence as a Consequence of pH and Storage Conditions, Library Quarterly, 39, 153 (1969).

29. Y.P. Kathpalia, Conservation and Restoration of Archive Materials, Unesco, Paris, 1973, Chapter 5.

30. W.K. Wilson, M.C. McKiel, J.L. Gear and R.H. MacClaren, Preparation of Solutions of Magnesium Bicarbonate for Deacidification, American Archivist, 41, 67 (1978).

31. Library of Congress Information Bulletin, 41, No. 13 (1982).

32. A.D. Baynes-Cope, The Non-Aqueous Deacidification of Documents, Restaurator, 1, 1 (1969).

33. R.D. Smith, Mass Deacidification: The Wei T'o Way, Research Libraries News, 45, 2 (1984).

34. J.M. Banks, Mass Deacidification at the National Library of Canada. Conservation Administration News, January 1985, No. 20; pp. 14-15; 27.

35. Y.P. Kathpalia, Deterioration and Conservation of Paper, IV. Neutralization, Indian Pulp and Paper, XVII, 230 (1962).

36. W.H. Langwell and J. R. Ede, Sulphur Dioxide and Vapour Phase
    Deacidification. 1967 London Conference on Museum Climatology,
    G. Thomson, London. International Institute for Conservation of
    historic and Artistic Work, 1968.
37. W.H. Langwell, The Vapour Phase Deacidification of Books and
    Documents. J. of the Society of Archivists, 3, 137 (1966).
38. W.H. Langwell, Technical Notes, The American Archivist, 29, 567
    (1966).
39. B.F. Walker, Morpholine Deacidification of Whole Books in
    Preservation of Paper and Textiles of Historic and Artistic
    Value (J.C. Williams, ed.), 1977, pp. 72-87.
40. G. Kelly, Non-Aqueous Deacidification of Books and Paper in
    Conservation of Library and Archive Materials and the Graphic
    Arts (G. Petherbridge, ed.), London, Butterworths, 1987, p. 117.
41. J.C. Williams and G.B. Kelly, Jr., Method of Deacidification,
    U.S. Patent 3,969,549 (1976); 4,051,276 (1977).
42. R.D. Smith, Mass Deacidification: the Wei T'o Understanding,
    C & RL News, January 1987, p. 2.
43. R.D. Smith, Preserving Cellulosic Materials Through Treatment
    with Alkyene Oxides, U.S. Patent 3,676,055 (1971).
44. R.A. Kundrot, Deacidification of Library Materials, U.S. Patent
    4,522,843 (1985).
45. W.H. Langwell, Prevention of Deterioration of Cellulose Based
    Records, U.S. Patent 3,472,611 (1969).
46. A. Rose and E. Rose, The Condensed Chemical Dictionary, 5 ed.,
    Reinhold Publishing Co., New York, 1956, p. 326.
47. G.V. Lomonova, Toxicity of Cyclohexylamine and Cyclohexylamine
    Carbonate, Prom. Toksikol. i. Kliniko Profe. Zabolevanii Khim,
    Etiol. Sbornik, Gos, Izd. Med. Lit., Moskva, 1962, pp. 160-163.
48. E.R. Plunkett, Handbook of Industrial Toxicology, Chemical Pub.
    Co., New York, 1966, p. 116.
49. I. Rosenblum and G. Rosenblum, Autonomic Stimulating and Direct
    Actions of Cyclohexylamine in Isolated Tissues, Toxicology and
    Applied Pharmacology, 13, 339 (1968).
50. V.V. Paustovskaya, M.B. Rappoport and M.F. Lyadenko, Material
    for a Toxicological Evaluation of Cyclohexylamine Chromate,
    Farmakologiya i Toksikologiya (Kiev), 2, 184 (1966).
51. G.V. Lomonova, Toxicity of Cyclohexylamine and
    Dicyclohexylamine, Federation Proceedings, translation
    Supplement, 24, 96 (1965).
52. The Merck Index: An Encyclopedia of Chemicals and Drugs
    (M. Windholz, ed.), 9th ed., Merck & Co., Inc., Rahway, N.J.
    p. 357.
53. R.H. Dreisbach, Handbook of Poisoning: Diagnosis Treatment, 5th
    ed., Lange Medical Pub., Los Altos, California, 1966, p. 121.
54. The Sigma-Aldrich Library of Chemical Safety Data (R.E. Lenga,
    ed.), Sigma-Aldrich Co., Milwaukee, WI, 1985, p. 520.
55. Encyclopedia Chimica, Kyoritsu Pub. Co., Tokyo, 1963, Vol. 9,
    p. 687.
56. G.S. Egerton, The Role of Hydrogen Peroxide in the Photochemical
    Degradation of Cotton Sensitized by Vat Dyes and Some Metallic
    Oxides, J. Text. Inst., T305-T318, September, 1948.

57. D.N.-S. Hon, Photochemical Degradation of Lignocellulosic Materials in Developments in Polymer Degradation - 3 (N. Grassie), Applied Sci. Pub., Essex, England, 1982.

58. R.A.V. Raff, R.D. Ziegler and M.F. Adams, Archives Document Preservation II, Northwest Sci., 41, 184 (1967).

59. O. Wachter, Konservierungstenchniken Fur Zeitungspapier, Austellung Der Osterreichischen Nationalbibliothek, Prunksaal, Juni-Oktober, 1984.

60. O. Wachter, Project Zeitungskonservierung, Osterreichische Nationalbibliothek Jahresbericht, 1984.

61. O. Wachter, Paper Strengthening: Mass Conservation of Unbound and Bound Newspapers. Paper presented at the Preservation of Library Materials Conference, Vienna, April 7-10, 1986.

62. C.C. Mollett, C.E. Butler and M.L. Burstall, Treatments of Archival Material, G.B. Patent 2,156,830A (1985); 2,180,248A (1987).

63. D.N.-S. Hon, Discoloration and Deterioration of Modern Papers, Science and Technology in the Service of Conservation, Preprints of the Contributions to the Washington Congress, September 3-9, 1982. pp. 89-92.

RECEIVED February 22, 1989

Chapter 3

Graft Polymerization

A Means of Strengthening Paper and Increasing the Life Expectancy of Cellulosic Archival Material

C. E. Butler[1], C. A. Millington[2], and D. W. G. Clements[3]

[1]Department of Chemistry, University of Surrey, Guildford, Surrey GU2 5XH, United Kingdom
[2]Department of Chemical and Process Engineering, University of Surrey, Guildford, Surrey GU2 5XH, United Kingdom
[3]Preservation Service, The British Library, Great Russell Street, London WC1B 3DG, United Kingdom

The rapid rate of deterioration of paper manufactured since the 1850's is of major concern to many libraries and custodians of cellulosic archival material. The main cause of this deterioration is the presence of acid which degrades the cellulose – the main constituent of paper. This acid is mainly derived from the use of alum–rosin sizing, alum is a particularly acidic substance. Conservators have approached this problem by developing methods which neutralise the existing acid and also leave a buffer against future acid attack. These methods undoubtedly extend the life of the paper, but do not unfortunately strengthen already weakened papers. An ideal solution would be to combine a neutralising process with one that strengthens weakened paper in a single operation. This would be a bulk treatment process in order to cope with the large backlog of essential conservation work. An approach to this problem has been under development for the British Library by the Industrial Chemistry group based at the University of Surrey. This article describes the techniques used and their application to whole book treatment. Some of the relevant points for a commercial-scale operation are also discussed.

The larger proportion of archival and modern information is stored on materials with a limited life expectancy. These materials include, paper, film, magnetic tape and optical disc. Paper, the major storage medium to date, is known to become brittle and discoloured with increasing age. This is particularly true of paper manufactured since the 1850's.

Libraries are the custodians of our heritage – a considerable amount of knowledge is held in these centres. However in recent years such establishments have become concerned with the increasing rate of deterioration of books and paper. One approach to solving

0097–6156/89/0410–0034$06.00/0

this problem has been to microfilm this information, but there is resistance to this because retrieving the information from screens is far less satisfactory than working from sheets of paper. The method of microfilming has its place in conservation as a means of saving the information contained in books which can no longer be used by readers, but still have sufficiently flexible pages to enable copying. Conservation reproduction methods include the use of overhead photocopiers or special photographic plates which can be slid between pages and then used to produce an image, but these methods are still relatively expensive. However there are also problems with books which have become too brittle for copying processes and a method which can restore sufficient strength to the paper to enable microfilming would be useful. Although Libraries accept that it is not always possible to preserve the written word in its original form (i.e. as a book or sheet of paper) an ideal conservation process would at least reduce the rate of deterioration of the paper sufficiently to enable continued use by readers and scholars for a number of years.

The importance of maintaining conservation programmes is reflected in the estimated value of library collections. For example the British Library's central collection is estimated to be worth in the order of £1,000 million; this does not include the large and valuable collection of manuscripts which is estimated to be worth a further £10,000 million. The present preservation backlog is some 2 million items and this figure is increasing rapidly year by year as more books require some form of conservation treatment. In financial terms the backlog represents a conservation cost of £150 million and this figure is increasing by some £3 million per annum. Therefore the need for a low priced bulk conservation process has been recognised by many institutions.

Paper deterioration is caused by two major reactions, (1) the acid-catalysed hydrolysis of the cellulose fibres (the main constituent of paper) which shortens the chain length of the cellulose polymer, and which is reflected as a loss in the paper's strength and flexibility, and (2) oxidation of the cellulose and other constituents which can lead to discolouration. These reactions can occur separately or simultaneously during the ageing of cellulosic materials. The main source of acid in paper comes from alum rosin sizing (alum is particularly acidic as it is the salt of a weak base and strong acid), other sources of acid include atmospheric pollution (sulphur dioxide is absorbed irreversibly and converted to acid); the use of iron gall inks (also containing sulphuric acid) and poorly purified pulp. Some of the break-down products of cellulose, lignin and other chemicals present in paper are also acidic and therefore could also act as a source of hydrogen ions for the hydrolysis reaction. The approach by conservators has been to neutralise (deacidify) the acid present in the paper and leave a buffer against future acid attack.

Many ways of deacidifying paper have been considered (1-5) and these include the use of calcium hydroxide/calcium bicarbonate solutions (Barrow's method); the use of magnesium bicarbonate in alcohols and other methods which to date have not been developed to any extent commercially. The main drawback with these methods are either the requirement to debind the book, or their labour intensive character; they are thus expensive. The only bulk deacidification process in commercial operation is based on the Wei T'o method (4) which uses either magnesium methoxide or methyl magnesium carbonate in a mixture of freons/methanol. This technique is being used by the Archives of Canada and a similar process has been installed by the French. Another bulk treatment which is under test for the Library of Congress is a vapour phase process using diethyl zinc (5). A recent review of the various preservation options is presented in a report by George Cunha (6) and in an article by the Congress of the United States Office of Technology Assessment (7). These articles do not discuss paper strengthening in any detail.

All the deacidification methods are effective at neutralising the acid present and buffering against further acid attack, and thus extending the life expectancy of the paper. They do not however generally restore strength to already weakened paper fibres. Estimates for the British Library suggest as much as 7-9% of their book stock consists of paper which is very brittle (i.e. 1 or 2 folds), while up to 25% of the stock of the Library of Congress is similarly affected. These figures represent a large proportion of books which require some form of conservation treatment in the very near future.

The strength of paper is normally assessed by its fold value (i.e. the number of folds before fracture) and a page with a fold value of less than 10 is generally considered brittle. Books with a low fold value would be particularly suitable for a paper strengthening process, especially if it enabled continued use of the books or rendered them suitable for copying processes. An ideal conservation process should aim to incorporate both neutralising and strengthening stages in one operation. An approach to this problem has been developed for the British Library by the Industrial Chemistry Group (ICG) based at the University of Surrey, Guildford, U.K.

PROCESS CRITERIA

Any process which is to be developed would have to fulfil certain criteria and these include:

(1) The rate of future deterioration should be reduced.
(2) The paper's mechanical properties should be enhanced.
(3) There should be no significant changes in the overall dimensions of the books.
(4) Changes in the appearance of the printed page should be minimised and should not affect use by readers.
(5) The process should be suitable for bulk treatment with the minimum of pre- and post treatment.
(6) The process should be cost effective.
(7) The process should be applicable to most types of paper.

GRAFT COPOLYMERISATION STUDIES

Early studies involved the development of a process which could treat all types of papers of varying ages and composition. Graft copolymerisation seemed to offer this potential. The basic concept of graft copolymerisation can be seen in Figure 1. Radical sites are created on the cellulose backbone and these sites allow the

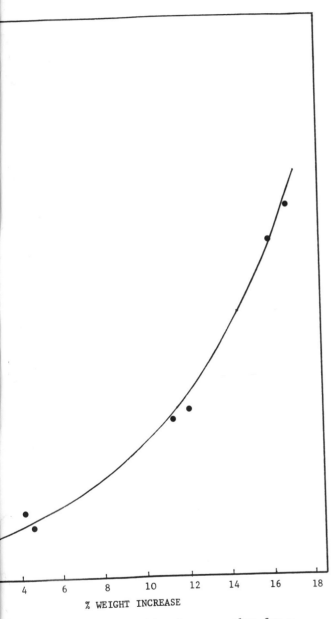

% Weight increase vs fold endurance value for a
paper using EA:MMA 5:1.

cellulose backbone

radical sites capable
of reacting with monomer

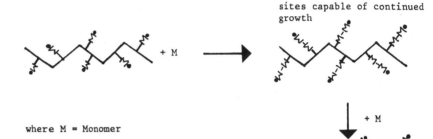

+ M

sites capable of continued
growth

+ M

where M = Monomer

+ M

etc. until the reaction is
completed or terminated

Representation by equations

Initiation   CellH $\longrightarrow$ Cell. + H.
Propagation   Cell. + M $\longrightarrow$ Cell-M.
              Cell-M. + M $\longrightarrow$ Cell-M-M. etc. until reaction is
                                               completed or
                                               terminated

Homopolymer formation

Initiation   M $\longrightarrow$ M.
Propagation   M. + M $\longrightarrow$ M-M. etc.
Termination M_x + M_y $\longrightarrow$ Inert products.

Where M = monomer and (.) represents a radical capable of being involved
in the polymerisation reaction.

Figure 1. A simplified representation of the grafting of
monomers onto the cellulose backbone.

attachment of monomer which is capable of continued polymerisation; thus the end result is a polymer chain which grows from the cellulose backbone.   Considerable work has been done on graft polymerisation, (8-12) and good reviews on cellulosic grafting are given by Arthur (8), and Heibeish and Guthrie (12).

The basic method used involves impregnating the cellulosic material, in our case – paper, with a monomer mixture and allowing an equilibration period to ensure that the monomer is homogeneously distributed throughout the paper before the polymerisation reaction is initiated.

The requirement of a bulk process implies that the treatment must be applicable to bound volumes and this therefore suggested the use of gamma rays to initiate the polymerisation reaction as these are able to penetrate books and are also capable of external control.

Early findings from this work were published in the Paris Conference proceedings of the IIC on Adhesives and Consolidants (13). This paper gave examples of the results obtained with different acrylic and methacrylic monomers. The monomers were selected because they had suitable physical characteristics and were already used in conservation work.   The problems which occur with aged and woody based papers were highlighted. This paper also indicated that the use of certain solvents such as methanol to improve monomer penetration had a detrimental effect on the strength of the paper, in particular its fold endurance.

Initial work used Whatmans filter paper (cotton based) as a control and Figure 2 shows that as the % polymer deposited within the paper structure increases the fold endurance also increases. Figure 3 shows a similar relationship for a lignin containing (wood based) paper, although the strength increases are usually less dramatic.

In general the papers which respond well to the process are those containing cotton fibres (Table I).   Increases in fold ratio (i.e.

Table I.   Fold Endurance (FE) for a variety of Different Paper Types and Ages Before and After Treatment

Date	Paper Type	% Weight increase	Initial FE	FE after Treatment
1838	Rag	20	80	1631
1839	Rag	24	13	119
1859	Rag	25	27	228
1865	Rag/mix	17	6	142
1874	Esparto	16	14	227
1890	Esparto	22	17	225
1903	Esparto	23	9	328
1874	Esparto/mix	25	18	48
1877	Esparto/mix	19	4	19
1949	Esparto/mix	21	10	39
1965	Mechanical	26	62	467
1969	Mechanical	18	103	1537
1970	Mechanical	20	25	146
1982	Chemical	25	175	3158

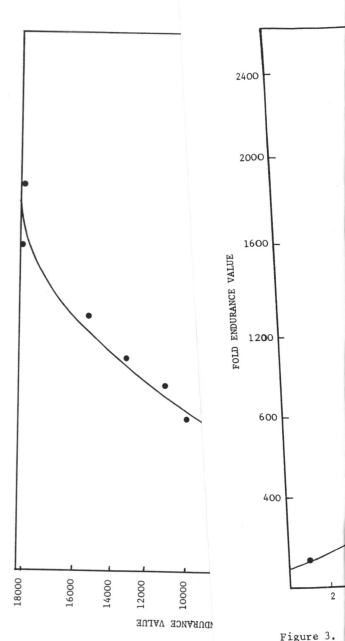

Figure 3. mechanical

the ratio of average fold after treatment/initial fold value) varies from 3-25 depending on the fibre type and the amount of polymer generated within the paper structure. Papers with low initial fold values tend to be more variable in their strength increases after treatment.

The rate of polymerisation is dependent upon the rate of formation of radicals within the paper structure. The use of high energy radiation enables the number of free radicals generated in a given time to be estimated. The polymerisation reaction is usually followed by studying changes in some physical parameter. In our work the % weight change and time were used to show the progress of the polymerisation reaction. This method enabled us to study factors which affect the rate of polymerisation; these included inhibitors which are substances capable of either retarding (slowing the rate of polymerisation), or preventing any polymerisation occurring. Other work was done to demonstrate the effect of various monomer combinations.

Studies showed that by varying the ratio of the monomers (in this case ethyl acrylate(EA) and methyl methacrylate(MMA), the shape of the kinetic curve could be altered. In effect the monomer mixture can be optimised to minimise the effect of inhibitors present in the system. Figure 4 shows the rate of polymerisation for different mixtures of EA:MMA incorporated into an esparto based paper. These curves can be compared to that of the individual monomers (Figure 5) which show that EA on its own gives very low polymer yields (i.e. the amount of polymer deposited within the paper compared to the target figure). Methyl methacrylate(MMA) gave a high polymer yield, however, it is a brittle polymer and therefore unsuitable for use on its own. The shape of the curve can also be influenced by different methacrylic monomers as shown in Figure 6. These types of curves have enabled us to acquire a clearer understanding of some of the factors which influence polymerisation reaction rates and polymer yields, and this understanding is important for any commercial scale application. Other factors which can affect the polymerisation rate include the gamma ray dose rate which affects the rate at which radicals are produced on the cellulose backbone, and in the monomer. Significant polymerisation will not occur in the presence of inhibitors unless they have first been consumed. Additional data are presented in 'Paper Conservator' (14) and in patents (15).

Some of the test work included the addition of a basic amino monomer which will neutralise much of the acid present in the paper prior to its incorporation into a polymer chain. The inclusion of the basic amino monomer and a multifunctional monomer (i.e. a monomer containing more than one double bond and is therefore capable of producing additional crosslinks) in the EA:MMA mixture gave increased strength gains relative to that of the basic monomer mixture. These additions led to the definition of a one stage deacidfying/strengthening process.

PROCESS SCALE UP

Once it had been established that the process was applicable to a large range of aged papers, work was undertaken to see if whole book treatment was possible. A simple process was used which involved the following steps:

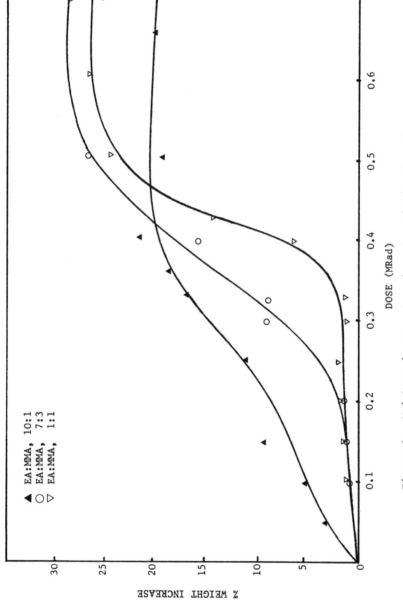

Figure 4. Weight gain as a function of radiation dose for an Esparto based paper with different ratios of EA:MMA.

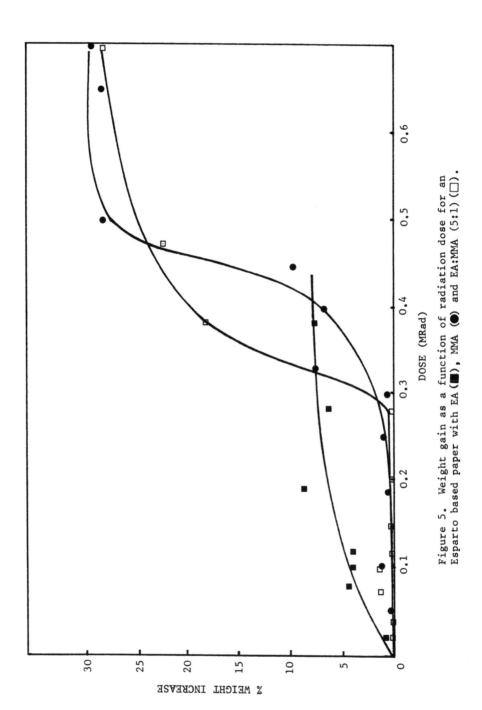

Figure 5. Weight gain as a function of radiation dose for an Esparto based paper with EA (■), MMA (●) and EA:MMA (5:1) (□).

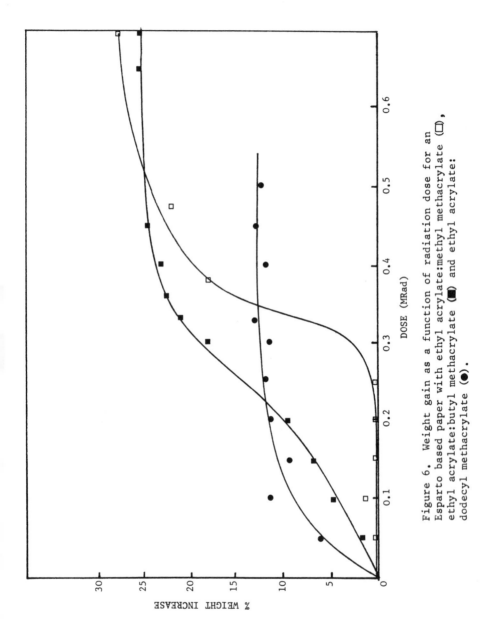

Figure 6. Weight gain as a function of radiation dose for an Esparto based paper with ethyl acrylate:methyl methacrylate (□), ethyl acrylate:butyl methacrylate (■) and ethyl acrylate: dodecyl methacrylate (●).

(a) Placing the books in a suitable container.
(b) Generating a suitable inert atmosphere within the book and the container and introducing the monomer.
(c) Allowing sufficient equilibration time to ensure the monomer was homogeneously distributed through the book and pages.
(d) Initiating the polymerisation reaction using low intensity gamma rays.
(e) Removing any residual odour prior to returning the books to the shelves.

Initial work concentrated on four main areas:

(1) Container design and construction.
(2) Monomer introduction and distribution.
(3) Process optimisation to ensure maximum strength gains and acidity reductions.
(4) Testing treated books with other conservation procedures.

The container was designed to hold between 5–10 books in any desired orientation. The size was initially based on a random selection of books taken from the Humanities and Social Science collections of the British Library. The container was built to withstand the calculated maximum and minimum pressure changes which were likely to occur during processing. Monomer introduction into the container took place under ambient conditions and several different procedures were developed and tested.

In order to monitor the changes in temperature and pressure which occurred during processing, thermocouples and a pressure transducer were fitted to the container. Figure 7 shows a typical temperature/time curve for a series of modern chemical paper based books, while Figure 8 shows the curve for a series of books of varying ages. The percentage weight gains are usually in the range 15–20%. These curves indicated that the overall reaction time and temperature rise were controllable. The actual scale for the time axis depends upon the irradiation dose rate used and the geometry of the treatment container. The optimum dose rate is such that the total irradiation dose received by the books is significantly less than that which would cause degradation of the cellulose, i.e. < 1.5 MRad.

The shape and size of the container which controls the rate of heat loss can be selected such that the time during which the books are subject to high temperatures can be kept small in order to minimise the effects of temperature induced cellulose degradation. It is anticipated that the maximum temperature in the full scale process will be less than 80°C.

Figure 9 shows a typical pressure curve and indicates that very low pressure rises occur during the polymerisation process.

The first series of trial runs established the following points:

(a) Consistent results were obtained with identical books.
(b) The monomer mixture(s) were capable of optimisation.
(c) The range of equilibration times was narrow; too short a time will cause uneven distribution, while excessively long times could promote the leaching of certain dyes and print toners.

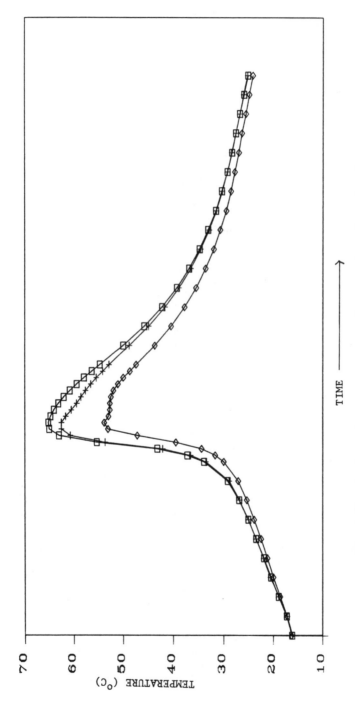

Figure 7. A plot of temperature vs time for the polymerisation
reaction with a series of modern chemical paper based books.

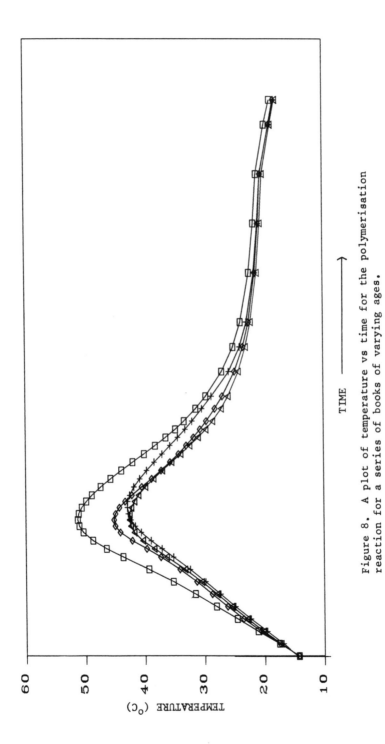

Figure 8. A plot of temperature vs time for the polymerisation
reaction for a series of books of varying ages.

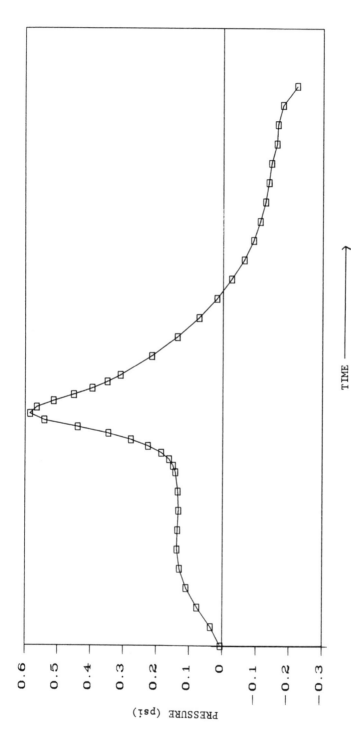

Figure 9. A plot of pressure vs time for the polymerisation reaction shown in Figure 7.

(d) The changes in pressure were low and this will facilitate the use of light containers.

Other points which were also established included:

(1) Good polymer yields could be obtained in the books.
(2) The between-book and the within book (page to page) distribution appeared acceptable.
(3) The distribution through the page was good.
(4) Low levels of residual monomer were observed, provided that complete polymerisation had occurred. Any residual odour could be removed by simple techniques.
(5) Strength increases were in line with those observed on the laboratory scale.
(6) Dyes and toners showed little or no movement provided excessively long equilibration times were avoided.
(7) Sewn bindings appeared undamaged.

COMMERCIAL SCALE CONSIDERATIONS

The design of a commercial-scale process has been initiated and the preliminary design studies and process costings have been completed. It is obvious that, if for instance 200,000 books per year are to be treated, the treatment plant must be designed as a continuous chemical process. The major cost item will be the gamma ray source and to ensure a cost efficient process this facility should be used continuously. Hence the definition of the number of separate processing stages and the time each book will take to move through each of these stages is important.
The process can be considered as follows:
(1) Book arrival at the process plant and movement to a suitable storage area.
(2) The preparation and loading of the treatment containers.
(3) The monomer introduction and distribution within each book in a container.
(4) The polymerisation in the gamma ray environment.
(5) The post irradiation treatment and movement of treated books to a transit store.

The attenuation of gamma rays by books depends upon the irradiation path length and the density of the book. In order to define the strength of the required gamma ray source and the size of the treatment container a statistical survey of the British Library's modern collection was undertaken. The histograms of height and mass of 18,000 randomly selected volumes are shown in Figures 10 and 11. These results indicate the wide variation which are likely to be encountered and together with data on the width and thickness have enabled the optimum methods of handling, packing and irradiation to be defined.
The costing exercise suggests that the price per book is at least comparable to other bulk treatment processes.

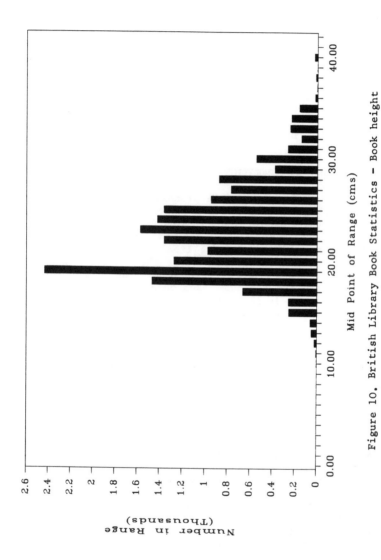

Figure 10. British Library Book Statistics – Book height

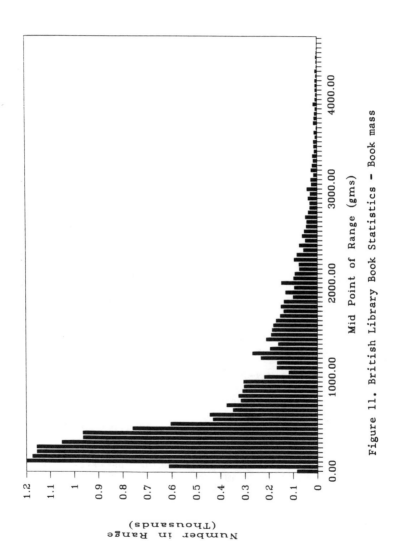

Figure 11. British Library Book Statistics – Book mass

STUDIES OF OTHER CONSERVATION PROCEDURES TO GRAFT COPOLYMERISED PAPER

Some preliminary investigations have been carried out on the graft
copolymerised paper to see which other conservation treatments could
be applied. This work included:

(1) The washing of the sheets in aqueous magnesium bicarbonate. The
    pH's of the untreated and treated paper were similar after this
    treatment.
(2) The use of spray deacidification using methyl magnesium carbonate
    which also gave satisfactory results.
(3) Repair methods using Japanese tissue and, wheat and starch
    pastes.
(4) Lamination with Paraloid and Texicryl adhesives.
(5) Studying the effects of accelerated aging.

ACKNOWLEDGEMENTS

This research was financed by the British Library. We would also
like to thank Dr. C. C. Mollett, Dr. M. L. Burstall, Mr. M. N. Nevin,
the British Library Steering Committee and all people who have worked
on this project.

REFERENCES

1. Barrow, W. J. Manuscripts and Documents: Their Deterioration and
   Restoration; Charlottesville, University of Virgina Press, 1965.
2. Barrow, W. J. Procedures and Equipment used in the Barrow Method
   of Restoring Manuscripts and Documents Richmond Va, 1965.
3. Langwell, W. H.; J. Soc. Archivists. 1973, 3(3),81–90.
   J. Soc. Archivists. 1979, 4(7), 597–598.
4. Smith, R. D.; U.S. Patent 3,676,055, 1972.
5. Williams. J. C. and Kelly G. B. Jr. U.S. Patent 3,969,549, 1976.
6. Cunha, G. M. Mass Deacidification for Libraries; Library
   Technology Reports May–June 1987, p 361–472.
7. Book Preservation Technologies, Congress of the United States
   Office of Technology Assessment. Washington DC 20510–8025; OTA-O-
   376, 1986.
8. Arthur, J. C. Graft Copolymerisation, Advances in Macromolecular
   Chemistry 1970, 2, 1–88.
9. Schwab, E., Stannett, V., Rakowitz, D. H. and Magrane, J. K.
   Paper Grafted with Vinyl Monomers using the Cerric Ion Method,
   Tappi, 1962, 45(5) p 390–400.
10. Koybayashi, A., Phillips, R. B., Brown, W. and Stannett, V.
    Tappi, 1971, 54(2), p 215–222.
11. Huang, R. Y-M and Rapson, W. H.; J. Polymer Sci. (2), 1963, p
    169–188.
12. Hebeish, A. and Guthrie, J. T., The Chemistry and Technology of
    Cellulosic Copolymers, Springer-Velag, N. Y. 1981.
13. Burstall, M. L., Mollett, C. C. and Butler, C. E., Graft
    Copolymerisation as a Method of Preserving Papers: Problems and
    Potentialities, Preprints of the Contributions to the
    International Institute of Conservation of Historic and Artistic
    Works (IIC), 1984, p 60–63.

14. Burstall, M. L., Butler, C. E. and Mollett, C. C. Improving the Properties of Paper by Graft Copolymerisation, <u>Paper Conservator</u>, Pt I, Vol. <u>10</u> 1986, p 95–100.
15. British Library. U.K. Patent GB 2,156,830, 1988.
   "   "   U. S. Patent 4,724,158, 1988
   "   "   U. S. Patent 4,808,433, 1989.

RECEIVED June 23, 1989

Chapter 4

Damaging Effects of Visible and Near-Ultraviolet Radiation on Paper

S. B. Lee[1], J. Bogaard, and R. L. Feller

Research Center on the Materials of the Artist and Conservator, Mellon Institute, Carnegie–Mellon University, 4400 Fifth Avenue, Pittsburgh, PA 15213

Little data have been available concerning chain breaking or an increase in the degree of oxidation of cellulose during exposure to the visible and near-ultraviolet radiation emitted by ordinary "daylight" fluorescent lamps under moderate conditions of temperature and humidity, both during exposure and during subsequent thermal degradation. The present investigation, involving papers of little or no lignin content -- an unbleached and bleached kraft pulp as well as filter paper -- revealed moderate immediate effects of exposure as well as sensitization towards subsequent thermal degradation. Intervention of an ultraviolet filter noticeably reduced, but did not prevent, deterioration both during exposure and during subsequent aging of 50% RH and 90°C. Continuous exposures of 800,000 to 1,300,000 footcandle hours were involved.

Although it is commonly stated that exposure of paper to light is potentially harmful, little information is available concerning how much and what kind of "harm" is done under conditions of moderate temperature, humidity and illumination. It was decided, therefore, to study the effects of exposure to "daylight" fluorescent lamplight, our particular source emitting only 3.8% near ultraviolet, and to express the results in terms of footcandle hours of exposure. Measurement of hot-alkali-soluble (HAS) matter would reflect the degree of oxidation of paper similar to the better-known measurement of copper number.([1]) In addition, changes in degree of polymerization (DP) were determined.

[1]Current address: Preservation Research and Testing Laboratory, Library of Congress, Washington, DC 20540

0097–6156/89/0410–0054$06.00/0
© 1989 American Chemical Society

Exposure can result thereafter in increased thermally-induced changes. In order to evaluate this aspect of the potential harm that exposure to light represents, the exposed sheets were subjected to thermal aging at 90°C and 50% RH and the changes in HAS matter and DP measured. Thus, not only were the immediate changes that took place during exposure monitored but the effect of exposure upon subsequent thermally-induced deterioration were determined as well. This is basically the same technique of investigation used by Launer and Wilson in 1943.(2)

EXPERIMENTAL

CHARACTERISTICS OF STOCK PULPS. The characteristics of the bleached (BP) and unbleached (UBP) kraft pulps used in these studies have been previously described.(3) Principally the BP pulp contained about 0.24% lignin and an initial solubility in hot 1%-sodium hydroxide of about 4.7% (HAS matter). The UBP pulp contained 4.4% lignin and 2.7% HAS matter. Whatman No. 42 filter paper, having only about 1% HAS matter, was chosen to represent a cellulose of high purity.

MEASUREMENTS OF PROPERTIES. Water-leaf handsheets at a basis weight of about 75 g/m^2 were prepared from the BP and UBP pulps using a Williams Standard Sheet Mould. Solubility in hot-1%-sodium hydroxide was determined by TAPPI method T212 om 83. Intrinsic viscosity in cuene (cupraethylenediamine solution) and, therefrom, the DP_v were determined according to ASTM Standard D1795.

EXPOSURE TO LIGHT SOURCES. As previously described(3) exposures took place under a bank of six General Electric 48-inch high-output "daylight" fluorescent lamps. The spectral power curve for such a lamp and the tolerance in its correlated color temperature has been published.(4). Spectral distribution data have also been provided by Harrison.(5) Measurements with a calibrated International Light IL700 research radiometer indicated that, of the total milliwatts of visible and near-ultraviolet irradiance, the lamps emitted about 3.8% in the near ultraviolet. A separate sensor head provided readings in terms of footcandles (1 footcandle = 10.67 lux). The lamps were mounted 3 1/2 inches above samples placed on a wire shelf in a room maintained at 50% RH and about 23°C. Owing to an unavoidable heating effect of the lamps, the test papers reached a temperature of about 29°C. Exposures were also carried out under a Plexiglas UF-3 sharp-cut-off filter (Rohm and Haas Company) which transmits less than 5% energy below 400 nm. Under the filter the luminous intensity was reduced about 10%; in the figures this correction has been made for the reduced effective time of exposure. Exposures were only to one side of the test sheets; previous studies indicated that the same net exposure on two sides would have much the same result.(6)

THERMAL AGING. Aging at 90°C and 50% RH took place in a humidity-controlled Blue M oven.

RESULTS

UNBLEACHED PULP. Initially, the development of hot-alkali-soluble
(HAS) matter in water-leaf handsheets prepared from unbleached pulp
(UBP) was followed both during exposure and during subsequent
thermal aging of the exposed sheets at 90°C and 50% RH. The
thermally-induced generation of HAS matter in the exposed test
sheets was compared to that which occurred in an unexposed control.
An increase in the degree or rate of generation of HAS matter
during thermal aging subsequent to exposure was taken as an
indication that potential damage had been done to the paper as the
result of the exposure.

As seen in Figure 1, the sheets, which had been exposed to a
total of about 800,000 footcandle hours, developed a greater amount
of HAS matter upon subsequent thermal aging than did the unexposed
control. Exposure to visible radiation alone (exposure under a
Plexiglas UF-3 ultraviolet filter) resulted in less generation of
HAS matter than when the sheets were exposed to unfiltered
radiation. If one may envision thermal deterioration of the papers
to occur in at least two principal stages(6), then the results
shown in Figure 1b suggest that exposure to light tends to increase
the extent of thermal degradation that occurs subsequently in an
initial stage or stages. Thereafter, as suggested by the dashed
lines, thermally-induced oxidation proceeds at a rate similar to
that experienced by the unexposed control.

If this interpretation of the results is valid, one may state
that exposure of the UBP sheets to "daylight" fluorescent lamplight
tends to cause some immediate photochemical damage, more oxidative
in character than directly causing chain scission.(7) This is
reflected in a rise in HAS matter during exposure and also during
an initial rapid stage of thermal deterioration. Thereafter, the
rate of generation of HAS matter at 90°C and 50% RH resembles
somewhat the rate exhibited by unexposed controls. With removal of
the ultraviolet portion of the irradiance from the "daylight"
fluorescent lamps by use of a Plexiglas UF-3 filter, the extent of
subsequent generation of HAS matter is reduced to about one-half
the rate without the filter. The immediate damage in this paper,
which contained about 0.24% lignin, was reduced, but not
eliminated, by removal of ultraviolet radiation.

BLEACHED PULP. The initial experiment was followed by similar
exposures of handsheets prepared from a bleached pulp (BP). Again
the UF-3-filtered situation resulted in a lesser development of HAS
matter (Figure 2a). The initial photochemical and subsequent
thermally-induced changes in HAS matter were less extensive than in
the case of the unbleached pulp (2a and b).

Owing to the very low lignin content in this particular pulp,
the viscosity in cuene solutions could be determined and from this
the fall in DP of the cellulose estimated (Figure 2c and d). In
Figure 3, a plot of the change in inverse of the degree of
polymerization (1/DP) with time indicates that measurable chain
breaking may have occurred even when the UF-3 filter was inter-
posed. The lower degree of photochemical damage under the UF-3
filter is also reflected in the subsequent rate of chain breaking

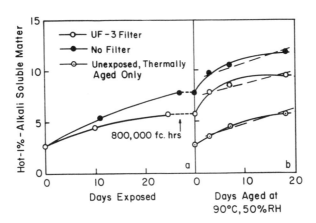

Figure 1. Development of hot-alkali-soluble matter in hand-sheets of unbleached pulp during exposure to "daylight" fluorescent lamps and subsequent thermal aging.

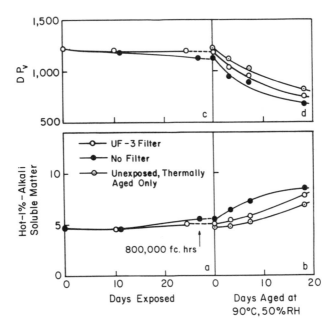

Figure 2. Development of hot-alkali-soluble matter and loss of degree of polymerization in bleached pulp during exposure to "daylight" fluorescent lamps and subsequent thermal aging.

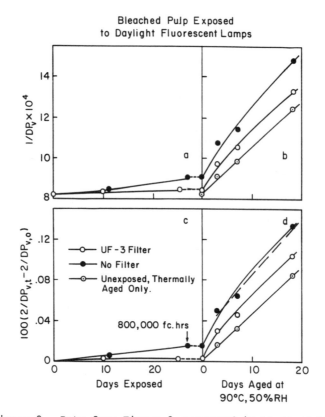

Figure 3. Data from Figure 2 expressed in terms of the
percentage of links broken, $100(2/DP_{v,t} - 2/DP_{v,o})$.

when the exposed samples were thermally aged at $90^{\circ}C$ and 50% RH (Figure 3d). As with the UBP, the exposed samples of BP seem to exhibit an increase in an initial rapid stage of thermally-induced chain breaking followed by a rate similar to that of the unexposed sheets; again, the dotted line is intended to indicate the similarity in rate after the initial rapid rise.

FILTER PAPER. Lastly, the same test procedures were carried out on Whatman No. 42 filter paper. Folding endurance tests were also performed to indicate in a practical way the immediate effect of exposure and to demonstrate how extensively the pulp had been degraded in the subsequent thermal-aging test (Figure 4f). The results again indicate an increased rate of development of HAS matter and also of chain breaking (Figure 5b) during an initial stage of thermally-induced aging followed by a stage in which the rate is similar to that of the unexposed control.

CHAIN BREAKING DURING EXTENSIVE CONTINUOUS EXPOSURE. Test sheets based on the bleached pulp (BP) stock and on Whatman No. 42 filter paper have been involved in a number of experiments at the Research Center under these same conditions of exposure. The results of four investigations involving the filter paper and two involving BP are summarized in Figure 6. The approximate percentage of links broken is reflected in the measurement $100[2/(DP)_{v,t} - 2/(DP)_{v,o}]$. We see that, in these two papers, having no more than 0.24% lignin in one case and none in the other, exposure to about 550,000 footcandle hours of illumination from the "daylight" fluorescent lamps resulted in the breaking of between 0.03 and 0.04% of the bonds in the filter paper and between 0.010 to 0.014% of the bonds in the BP. If it requires the breakage of between 0.3 and 0.6% of the bonds in cellulose to reduce paper to practically zero folding endurance (see Figures 4f and 5b and also reference 6), and if one assumes that a linear rate of chain breaking takes place, one may calculate a minimum "lifetime" before all folding endurance is lost to be about 5.5 million footcandles of exposure for the one paper and about 15.4 million footcandle hours for the other. Although these are only rough estimates, the results suggest, in the worst case, that we are talking about something on the order of 50 years' exposure on a gallery wall, well-illuminated by diffuse daylight at an average level of about 30 to 38 footcandles.(8) If a moderate level of illumination of about 10 footcandles is employed in the display of works on paper, the estimate becomes well over 100 years for these papers having little or no content of lignin. For papers containing considerable lignin, the potential for damage, of course, is much more significant.(2)

These preliminary studies to evaluate the potentially damaging effects of the direct exposure to test sheets to visible and near-ultraviolet radiation employed exposures tens of times greater than those customarily used in "light bleaching" procedures. Moreover, the test sheets have been exposed directly to the radiation in contrast to the customary exposure of the papers in an aqueous bath to achieve bleaching. Experiments to be reported elsewhere show that the damage to the paper is very much reduced when the sheets are immersed at less than a centimeter

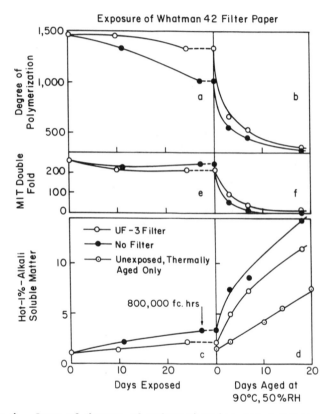

Figure 4. Loss of degree of polymerization and folding endurance and rise in hot-alkali-soluble matter in filter paper during exposure to "daylight" fluorescent lamps and subsequent thermal aging.

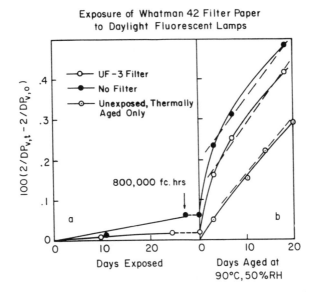

Figure 5. Data on loss of DP from Figure 4 expressed in terms of the percentage of links broken, $100(2/DP_{v,t} - 2/DP_{v,o})$.

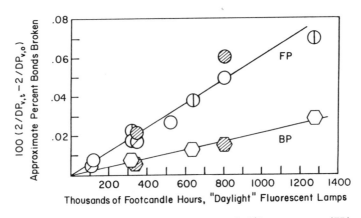

Figure 6. Data on the direct exposure of filter paper (FP) and bleached pulp (BP) summarizing the results in terms of the percentage of links broken versus footcandle hours of exposure to "daylight" fluorescent lamps. Each set of symbols signifies a different experimental run; the shaded data points refer to the present results shown in Figures 3 and 5.

depth in water or in aqueous alkaline buffer solutions of about pH 8 (Feller, R. L., Lee, S. B. and Bogaard, J., Symposium 88, Ottawa, Canada, October 1988, to be published).

CONCLUSIONS

The results provide an indication of the potential rate of deterioration of papers of high or reasonably high quality in terms of footcandle hours of exposure to "daylight" fluorescent lamps under moderate conditions of temperature and humidity. The use of an ultraviolet filter considerably reduces but does not eliminate the harmful effects of the exposure. Papers having little or no lignin content appear to possess considerable resistance to damage by an essentially visible-irradiance source even when several percent near-ultraviolet is present. It is notable that no significant induction time is apparent; photochemical damage began immediately.
     A certain amount of chain breaking occurs during exposure, evidenced by a loss in DP and possibly by the increase in HAS matter. This is the "immediately harmful" effect of exposure. Some oxidation of the cellulose chains also seems to take place, suggested by the rise in HAS matter during exposure and by the initially rapid rate of change that occurred upon subsequent thermal degradation of the exposed papers. This represents the "potentially harmful" effect of exposure, leading to loss in degree of polymerization through thermally-induced reactions. These concepts are not new but the authors trust that a clear demonstration of the effects has been useful.

LITERATURE CITED
1. Feller, R. L.; Lee, S. B.; Curran, M. Three Fundamental Aspects of Cellulose Deterioration; Supplement, Art and Archeology Technical Abstracts, 1985, 22, No. 1, pp. 329-354.
2. Launer, H. F. and Wilson, W. K. "Photochemical Stability of Papers", J. Res. Natl. Bur. Stand., 1943, 30, 55-74.
3. Lee, S. B.; Feller, R. L. "Influence of the Hemicellulose Fraction on Thermal and Photochemical Discoloration of Paper", in Adv. Chem. Ser. 212; Needles, H. L. and Zeronian, S. H., Eds.; 1986, 377-386.
4. IES Lighting Handbook, Reference Volume, Illuminating Engineering Society; New York, 1981, Figures 8-21 and 8-36.
5. Harrison, L. S. Report on the Deteriorating Effects of Modern Light Sources, Metropolitan Museum of Art, New York, 1955.
6. Feller, R. L.; Lee, S. B.; Bogaard, J. In Adv. in Chem. Ser. 212; Needles, H. L. and Zeronian, S. H., Eds.; 1986, 329-346.
7. Lee, S. B.; Feller, R. L.; Bogaard, J. J. Imaging Sci., 1985, 29, No. 2, 61-64.
8. Feller, R. L. Museum, 1964, 17, 57-98.

RECEIVED February 22, 1989

Chapter 5

The Effect of Variations in Relative Humidity on the Accelerated Aging of Paper

Chandru J. Shahani, Frank H. Hengemihle, and Norman Weberg

Preservation Research and Testing Office, Library of Congress, Washington, DC 20540

The effect of fluctuations in relative humidity on the degradation of paper-based materials under accelerated aging conditions has been studied to gain a better understanding of environmental needs for long-term storage of paper-based archival materials. The rate of degradation of paper samples aged under a relative humidity ramping up and down between 40 and 60% every 12 hours at a constant temperature of 90°C, has been compared with corresponding data obtained under constant relative humidity conditions of 40, 50 and 60% at the same temperature. Two bleached Kraft papers, an unsized waterleaf and an alum-rosin-sized paper, were each aged as single sheets and in piles between Plexiglas sheets to simulate the book form. For single sheets aged under cycling humidity conditions, fold endurance, brightness and pH values declined at least as rapidly as under constant aging conditions of 90°C and 60% relative humidity. For samples aged as simulated books, the damping effect due to the increased paper mass was helpful in slowing down deterioration induced by the cycling humidity conditions. However, test samples inside simulated books invariably aged at a faster rate than single sheets, due probably to adverse conditions created within the book structure by trapped degradation products.

All too often, the environment within library and archive storage areas presents a threat to their aging collections, when it could be utilized as a most cost-effective tool for their preservation. All paper-based materials, whether rare or common-place, old or new, acid or alkaline, are susceptible to heat and humidity. While it is

0097–6156/89/0410–0063$06.00/0
© 1989 American Chemical Society

generally accepted that lower temperatures and lower levels of relative humidity are conducive to the preservation of archival materials, there is no consensus about the most desirable environmental conditions in areas that house such materials for long-term storage. In part, this is due to genuine differences in institutional goals and resources. But a lack of definitive experimental data has also contributed to the indecisiveness prevalent at this time.

One area where we need a better understanding is the effect of cyclic or repeated fluctuations in relative humidity on the aging of paper-based collections. Over the past several decades, we have so emphasized the importance of a steady temperature and relative humidity (RH), that a vacillating RH line on a hygrothermograph chart can cause much consternation and evince visions of crumbling collections. It is not surprising therefore, that when participating in the design of storage facilities, archivists and librarians feel secure with only the most stringent environmental controls.

Our fear of cycling environmental conditions probably has its root in museum conservation, which must care for delicate objects, often constructed of a variety of composite materials. Individual components of such materials may respond quite differently to changes in temperature and relative humidity, thereby leading to stresses which can be disastrous. However, if there were any doubts about the relevance of borrowed environmental control principles in archives and libraries, they were probably put to rest when Cardwell showed that paper aged much more rapidly when subjected to cycling environmental conditions than when aged at a constant temperature and relative humidity (1). The paper samples in these experiments cycled between a dry aging oven at 100°C, and ambient conditions of 23°C and 50%RH, every 24 hours. Cardwell obtained further confirmation of these data from experiments in which test papers conditioned at different relative humidities were sealed within glass tubes before aging them under cycling temperature conditions.

In the real world of libraries and archives, we are concerned with books, documents stored within boxes, or maps and prints inside drawers and cabinets — none of which are as open and accessible to environmental changes as a test sheet suspended in a laboratory oven. Also, the gyrations in environmental conditions in library and archive storage areas would not be nearly as extreme and violent as in Cardwell's study. One would intuitively expect any fluctuations in environmental humidity and temperature to translate into slower and smaller changes in the moisture content of a large body of paper which generally makes up an appreciable fraction of the total volume of the room.

Since it is relatively easier and cheaper to control temperature than relative humidity in storage areas, the immediate need is for a better understanding of the effects of cycling relative humidity conditions. Therefore, we have confined this study to a constant temperature of 90°C, while cycling the relative humidity between 40 and 60 per cent. One full cycle is completed every 24 hours. As a gross approximation, 3 days of accelerated aging at 90°C and 50% RH are generally equated to 20 to 30 years of natural aging under ambient conditions. In effect, the environmental fluctuations employed in this work are in all probability much

milder than those to which an average library book is subjected over
its lifetime. The observation of a measurable effect on the
physical properties of paper under these conditions suggests a need
for further investigation.

Experimental

Materials. All paper samples were cut from a continuous length of
machine-made rolls. Two different papers have been studied. One is
a bleached Kraft wood pulp waterleaf (50-1b basis weight) made by
Neenah Paper Mills from a stock that comprised northern softwoods
(60%) and Lake States hard woods (40%), and contained no additives.
The second paper, Foldur Kraft, is a bleached Kraft paper (70-1b
basis weight) made by Champion Paper Company from a stock of 90%
softwoods and 10% hardwoods, with alum-rosin size and 8% titanium
dioxide filler. This paper was obtained 15 years ago. Since then
it has been stored indoors in areas which have not enjoyed a
carefully regulated environmental control.

Aging Conditions. Paper Samples were aged as (a) loose sheets hung
vertically on a rack which permitted free air-flow around them, and
(b) 100 sheet piles placed between Plexiglas sheets. Test samples
were removed from the middle of the piles.
A Thermotron programmable humid oven was employed for the
accelerated aging experiments. The temperature of the oven was held
steadily at $90^{\pm}0.2^{o}C$ throughout this series of experiments. The
relative humidity in the cycling mode was fixed alternately at 40
and 60% for 11-hour intervals with intervening periods of 1 hour
each, during which it ramped up to 60% or down to 40%. Samples were
also aged under constant relative humidity conditions of 40, 50 and
60% at $90^{o}C$. The relative humidity conditions did not deviate more
than $\pm2\%$ from the programmed value.

Testing. Aged samples were conditioned at $23^{o}C$ and 50%RH for at
least 24 hours before testing was initiated. MIT Fold endurance was
determined along the machine direction in accordance with TAPPI
Standard Test Method T511, with a modification which reduced the
tension load from 1.0 kg to 0.5 kg. The number of double folds to
failure were measured. The pH of the paper was determined by a
cold-extraction technique based upon the TAPPI Standard Method T509.
A slurry containing 1.0 g of paper in distilled water was made up to
100 mL. The paper sample was macerated in a Waring blender for 1
minute. The slurry was allowed to stand for 5 minutes before its pH
was measured. Brightness levels were determined by measuring blue
reflectance with a Photovolt model 670 reflection meter.

RESULTS & DISCUSSION

Of the two test papers employed in this study, the wood pulp
waterleaf better represents natural cellulose, while Foldur Kraft
paper, which is sized with alum-rosin, is fifteen years old and also
fairly acidic, more closely resembles average library book paper.
Both the papers were aged as sheets freely suspended on a rack, as

is general laboratory practice in such experiments, and also in piles between Plexiglas sheets to simulate the book form.

The decrease in fold endurance along machine direction upon accelerated aging at 90°C for wood pulp waterleaf sheets and simulated books is shown in Figures 1 and 2, respectively. The corresponding data for Foldur Kraft paper are presented in Figures 3 and 4. The experimental points in these figures represent an average of at least 10 individual measurements, and the error bars indicate 95% confidence limits. The degradation of the two test papers under the different sets of accelerated aging conditions employed here, was also followed by pH and brightness measurements. These data are presented in Tables II through VII. The decline in pH values and the loss in brightness generally parallel the loss in fold endurance with progressive aging. Therefore, while the following discussion may neglect further reference to the pH and brightness data, these data do confirm the general trends gleaned from an examination of the fold endurance data.

Relative Lifetime: To facilitate a relative evaluation of the fold endurance data in Figures 1 through 4, we have arbitrarily selected a single point at which to compare all of the data sets. The time taken for the fold endurance value to decrease to an eighth of its initial value, termed as the "Relative Lifetime", has been computed for each of the least squares-fitted curves in the Figures. These values, which are presented in Table I, serve to give an overview of the fold endurance data. Inferences drawn from a comparison of these Relative Lifetime values should hold equally true under comparison at any other point of reference, or if drawn directly from the fold endurance loss curves in Figures 1 through 4.

Effect of increasing relative humidity: For samples aged under constant relative humidity conditions of 40, 50 and 60 percent, the declining fold endurance values in Figures 1 through 4 and Table I, present a familiar trend (5). The rate of loss in fold endurance accelerates with increasing relative humidity leading to proportionately smaller Relative Lifetime values in Table I. The observed trend is identical for samples aged as loose sheets and those aged as simulated books. Samples aged at lower relative humidity levels would have a smaller moisture content, and therefore a less swollen matrix, which would be less accessible to acids and oxidants. Hence the observed increase in stability with decreasing relative humidity. The effect of moisture content of paper on its "accessibility", and therefore, its reactivity, has received considerable attention in literature (6).

Accelerated aging of loose sheets and books: A comparison of the fold endurance data for loose sheets with the corresponding data for simulated books shows that for both the test papers, the paper within the book ages at an appreciably faster rate. A similar acceleration in the rate of degradation has been observed when paper sealed within polyester envelopes or glass tubes is subjected to accelerated aging (7). In such cases, it has been observed that the introduction of another sheet containing an alkaline reserve in the same capsule, nullifies the acceleration in the rate of degradation

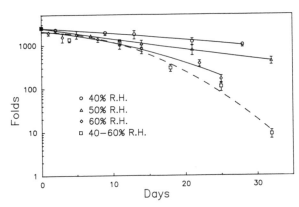

Figure 1: Effect of relative humidity on fold endurance of wood pulp waterleaf sheets aged at 90°C.

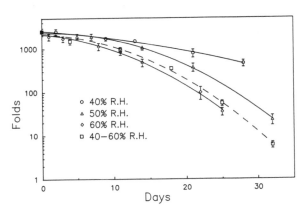

Figure 2: Effect of relative humidity on fold endurance of wood pulp waterleaf books aged at 90°C.

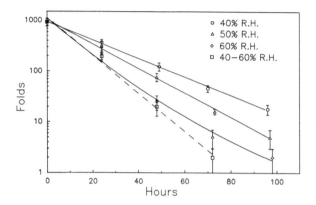

Figure 3: Effect of relative humidity on fold endurance of
Foldur Kraft sheets aged at 90°C.

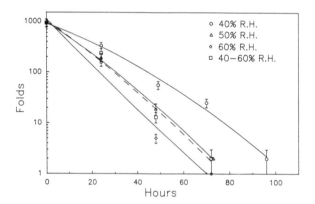

Figure 4: Effect of relative humidity on fold endurance of
Foldur Kraft books aged at 90°C.

TABLE I

Effect of humidity on Relative Lifetime values
of paper samples under accelerated aging at 90°C

Sample Arrangement	Relative Humidity (%)	Relative Lifetime*	
		Wood pulp waterleaf (days)	Foldur Kraft (hours)
S H E E T	40	49.8	51.0
	50	30.1	41.0
	60	21.1	28.3
	40 - 60	18.0	27.1
B O O K	40	31.2	39.2
	50	20.9	26.9
	60	16.1	20.0
	40 - 60	17.2	25.6

* Relative Lifetime = Time required for fold endurance to
  decrease to an eighth of initial value.

TABLE II

Effect of relative humidity on pH of wood pulp
waterleaf sheets aged at 90°C

RH(%)	40	50	60	40–60
Days		pH		
0	6.1	6.1	6.1	6.1
1	--	--	5.7	--
2	6.0	--	--	--
3	--	--	5.7	--
4	--	--	--	5.4
5	--	5.9	--	--
7	--	5.8	--	--
8	--	--	5.6	--
9	5.8	--	--	--
11	--	--	5.5	5.4
13	5.6	--	--	--
14	--	5.5	5.4	--
18	--	--	--	5.19
21	5.4	5.4	--	--
22	--	--	5.2	--
25	--	--	5.2	5.1
28	5.3	--	--	--
32	--	5.4	--	5.0
39	--	--	--	5.0
46	--	--	--	4.9

TABLE III

Effect of relative humidity on pH of wood pulp
waterleaf books aged at 90°C

RH(%)	40	50	60	40–60
Days		pH		
0	6.1	6.1	6.1	6.1
1	--	--	6.0	--
2	6.0	5.8	--	--
3	--	--	5.9	--
4	--	--	--	5.5
7	--	5.7	--	--
8	--	--	5.6	--
9	5.6	--	--	--
11	--	--	5.4	5.3
13	5.3	--	--	--
14	--	5.4	5.4	--
18	--	--	--	5.0
21	5.1	5.3	--	--
22	--	--	5.1	--
25	--	--	5.0	5.0
28	5.0	--	--	--
32	--	5.0	--	4.8
39	--	--	--	4.8
46	--	--	--	4.7

TABLE IV

Effect of relative humidity on pH of Foldur Kraft paper
aged as sheets and books at 90°C

RH (%)	40	50	60	40-60
Hours		pH		
Sheet	0	4.5	4.5	4.5	4.5
Sheet	24	4.3	4.5	4.2	4.3
Sheet	49	4.3	4.3	4.2	4.3
Sheet	70	4.2	4.2	4.2	4.3
Sheet	96	4.5	4.4	4.3	4.3
Book	0	4.5	4.5	4.5	4.5
Book	24	4.3	4.3	4.3	4.3
Book	49	4.2	4.2	4.2	4.2
Book	70	4.2	4.2	4.1	4.3
Book	96	4.2	4.1	4.1	4.2

TABLE V

Effect of relative humidity on brightness
of wood pulp waterleaf sheets aged at 90°C

RH(%)	40	50	60	40-60
Days		Reflectance (%)		
0	89.6	89.6	89.6	89.6
1	--	--	85.0	--
2	87.0	--	--	--
3	--	--	83.0	--
4	--	--	--	86.0
5	--	83.5	--	--
7	--	82.5	--	--
8	--	--	79.0	--
9	83.0	--	--	--
11	--	--	77.0	78.0
13	82.0	--	--	--
14	--	79.2	74.0	--
18	--	--	--	73.0
21	81.0	76.2	--	--
22	--	--	69.0	--
25	--	--	--	68.0
28	78.0	75.2	68.0	--
32	--	--	--	65.0
39	--	--	--	62.0
46	--	--	--	59.0

TABLE VI

Effect of relative humidity on brightness of
wood pulp waterleaf books aged at 90°C

RH(%)	40	50	60	40-60
Days		Reflectance(%)		
0	89.6	89.6	89.6	89.6
1	--	--	85.0	--
2	86.0	--	--	--
3	--	--	--	--
4	--	--	84.0	83.0
5	--	83.5	--	--
7	--	79.3	--	--
8	--	--	75.0	--
9	81.0	--	--	--
11	--	--	73.0	74.0
13	78.0	--	--	--
14	--	75.0	70.0	--
18	--	--	--	67.0
21	75.0	70.0	--	--
22	--	--	62.5	--
25	--	--	62.0	62.0
28	71.0	--	--	--
32	--	65.1	--	57.0
39	--	--	--	53.0
46	--	--	--	52.0

TABLE VII

Effect of relative humidity on brightness of Foldur Kraft
paper aged as sheets and books at 90°C

	RH (%)	40	50	60	40-60
	Hours		Reflectance (%)		
Sheet	0	75.5	75.5	75.5	75.5
Sheet	24	74.0	73.2	71.0	70.8
Sheet	49	72.0	71.0	69.0	69.2
Sheet	70	71.0	69.0	67.3	67.3
Sheet	96	71.0	69.0	66.6	66.4
Book	0	75.5	75.5	75.5	75.5
Book	24	73.5	71.9	70.0	70.7
Book	49	71.5	69.2	66.7	67.5
Book	70	68.0	67.4	64.2	65.6
Book	96	69.0	65.8	62.7	63.0

due to encapsulation (8). It may be that acidic degradation
products get trapped inside a polyester capsule, or within a book,
where they continue to accumulate, and thereby create an increasingly acidic environment inside the sealed paper.

Trends discerned from studies of deterioration patterns in
archival records appear to lend support to the observed acceleration
in the rate of degradation within air-tight envelopes and books.
One of the present authors conducted a survey of the condition of
pre-1840 records at the National Archive and Records Administration's Regional Archive in Philadelphia, PA (9). Loose papers
within boxes (all of which were constructed of highly acidic stock)
were found to be in remarkably good condition, while the paper in
bound records was generally weaker, and had a higher acid content.
A recent report from the National Research Council, commenting on
the condition of records at the National Archives and Records
Administration, Washington, D.C., also reports that paper records
stored within boxes were generally in better condition than those
within bound volumes (10). This report ascribes the better condition of paper within boxes to the more stable "microenvironment"
they provide by damping the temperature and humidity fluctuations
and by presenting a barrier against pollutant intrusion. On the
other hand, bound volumes, which were frequently discolored at the
edges, were viewed as being open to pollutants. On the other hand,
the fact that the discoloration stops at the outer edge, suggests
that a tightly closed book provides an effective barrier to penetration by pollutants. Also, most of the document storage boxes in use
at NARA until the late seventies, were constructed of highly acidic
stock and had finger-holes in their sides. It would appear therefore, that these boxes might not have provided as effective a
protection from the environment as the book form, but they did
provide an outlet for degradation products.

The observation of a faster rate of degradation within a book
as compared to a loose sheet under the same accelerated aging
conditions raises some thought-provoking questions. To the best of
our knowledge, in accelerated aging experiments with paper, the
general practice has been to employ loose sheets of test samples
suspended within an aging oven. If the rates obtained with such an
experimental set-up are different from those observed within a book,
can such tests truly project lifetime patterns for book paper?
Probably, the accepted methodology for accelerated aging tests needs
to be reexamined.

Effect of cycling relative humidity: Further examination of the
accelerated aging data in Figures 1 and 3 shows that under cycling
humidity conditions, loose sheets of both the test papers age at
least as rapidly as under a constant relative humidity of 60
percent. The adverse effect of cycling humidity conditions observed
here is not as striking as that observed by Cardwell (1) under a
much wider humidity and temperature cycle. However, given the small
range of the relative humidity cycle, the long cycle which trans-
lates to 6 to 8 years under ambient conditions, and the constant
temperature in the present experiments, the observed decrease in the
rate of degradation is remarkable. The data in Figure 2 for wood
pulp waterleaf aged in simulated book form, show that even with the

damping of humidity change within a book, the test papers still age
faster under a cycling RH, than at the median relative humidity of
the cycle. However, the rate of degradation within the book is
slower than that observed at a constant relative humidity of 60 per
cent. Figure 4 shows that for Foldur Kraft paper, the rate of loss
of fold endurance under cycling RH conditions is of the same order
as that observed at the median RH of 50 percent. This observation
confirms the fact that a larger body of paper, as a book, is not as
easily affected as a loose sheet of paper. The presence of the
sizing in this paper probably makes it less vulnerable to changes in
relative humidity than the unsized waterleaf. However, it is hard
to draw a definitive inference from a comparison of only two papers.
Nevertheless, it has been demonstrated here that constantly fluc-
tuating relative humidity conditions -- even when cycling between a
fairly narrow range -- do have the potential to accelerate the aging
process of paper-based materials. But further work is needed before
acceptable tolerance levels can be suggested.

Possible causes of degradation under cycling relative humidity: The
interaction of cellulose with water has been the subject of extensive study over several decades (2). However, we have failed to
uncover any investigations of moisture sorption and desorption that
were confined to an RH region as narrow as that studied in the
present work. Repeated wetting and drying cycles have been employed
to minimize the effect of changes in environmental humidity on the
moisture content of cellulose (3,4). It is interesting to consider
if such a "stabilization" process has any relevance in the present
context. The reduced change in the moisture content of cellulose
due to changes in environmental humidity as a result of repeated
moisture sorption-desorption cycles, has been reported to be
accompanied by a small increase in the crystallinity of cellulose
(11). Jeffries, who observed that the sorption of water by viscose
film decreased by as much as 60 percent after just six cycles
between 0 and 100 percent RH at $90^{O}C$, suggested that a small change
in crystallinity could not, by itself, explain the magnitude of the
stabilization effect (3). He ascribed the reduced ability of the
chain networks to expand and swell to "a cross-linking of the
amorphous material by a small amount of crystalline material
distributed throughout the structure." Preston et. al (12) have
also ascribed the stabilization phenomenon to a cross-linking action
of very small crystallites. It should be noted that Jeffries (3),
who had stabilized his samples between 0 and 100 percent RH at $90^{O}C$
for 3 to 5 weeks, recognized that the stabilization effect he
observed was partially caused by an increase in crystallinity due to
acid hydrolysis. Other workers have suggested that stabilization
results from a relaxation of internal strains within the fibrous
network (4). All of the suggested possibilities are in agreement
that stabilization of the moisture sorption properties of paper is
caused by physical changes in the morphology of cellulose. No
chemical change or reaction is believed to be involved in this
process.

An essentially physical change in cellulose, which can effectively reduce its moisture content, can indeed account for the
observed increase in fold endurance loss under cyclic RH conditions,

since the fold strength of paper has been shown to decrease with
decreasing moisture content (13). However, the loss in fold
endurance shown in Figures 1 through 4, is paralleled by a loss in
brightness and a decrease in pH values. These observations suggest
that a chemical interaction must be largely responsible. An
increase in the reactivity of the cellulose matrix can only be
explained by a more swollen and accessible matrix, and therefore, a
higher moisture content. Even though paper has not been subjected
to extremes of dryness and humidity in the present work, as is
generally the practice in stabilization experiments, there can be no
justification for a contrary trend in moisture absorption. This
apparent dichotomy is most satisfactorily reconciled by the work of
Brickman et al. (14), who subjected dewaxed cotton linters to
repeated wetting and drying cycles. They observed that even though
the moisture regain and the heat of hydration of cotton linters were
reduced, the reactivity, and hence accessibility, actually increased.

Some of the earliest evidence of increased accessibility of
cellulose as a result of repeated wetting–drying cycles was
presented by Rosenthal and Brown (15), who observed an increase in
the extent of nitration. Brickman et al. (14) extended this work,
and confirmed enhanced reactivity towards nitration and thallous
ethylate in ether, besides an increment in hydrogen–deuterium
exchange. Subsequently, Yin and Brown (16) reported a similar
increase in reactivity when the wetting step was performed at 0 or
$25^{o}C$. However, they observed a decrease in reactivity for samples
wetted at or above $50^{o}C$. In this context, the present observation
of increased reactivity under cycling humidity conditions at $90^{o}C$,
is in contrast with their observation. The two systems, of course,
differ greatly in detail, especially in experimental conditions
during the moisture sorption–desorption step.

Several possibilities present themselves when one attempts to
define a probable cause for the observed increase in the rate of
degradation under cycling humidity conditions. However, in the
absence of any data on moisture sorption and desorption, all such
considerations can only be speculative. It may be that the so-
called junction points (17), which represent inaccessible groups of
relatively ordered molecules within amorphous regions, are somehow
forced open by the repetitive surge in water concentration. A more
plausible possibility is that the constant flux of water molecules
in and out of the fibrous network may facilitate the hydration of
protons on acid molecules, thereby increasing their mobility and
therefore, the probability of their interaction with cellulosic
hydroxyl groups. Oxidative degradation of cellulose is also
facilitated by water molecules (18), so an increased flux of water
molecules through the cellulose matrix may also increase the rate of
such reactions. Clearly, more work is needed to comprehend reaction
mechanisms underlying the observed degradative effect under cycling
humidity conditions.

Conclusion: For the librarian and the archivist, who must operate
from an essentially practical concern for the storage of archival
materials, the stabilization of moisture sorption properties of
paper as a result of humidity cycling and mechanistic considerations

at the molecular level appear to be of little significance at this point. For them, the inescapable conclusion is that when cellulose is subjected to cyclic humidity conditions, it does become more accessible to chemical reactants, causing paper to age at a faster rate. However, there is some solace to be derived from the damping effect provided by the book structure. Here too, further work is needed to quantify the interaction between a large body of paper and the environment. While the book may offer some hope for damping fluctuations in environmental conditions, a new question has been raised about the possibility of trapping of acidic degradation products within the book.

Literature Cited

1. Cardwell, R. D., Aging of Paper, Ph.D. Thesis, N.Y. State College of Forestry, Syracuse, N.Y., 1973, pp. A1-8.

2. Zeronian, S. H. Intercrystalline swelling of cellulose in Cellulose chemistry and its applications; Nevell, T. P.; Zeronian, S. H., Eds.; John Wiley & Sons: New York, 1985, pp. 138-158.

3. Jeffries, R. J. Textile Inst. 1960, 51, T339.

4. Zeronian, S. H.; Kim, M. S. Proc. Tenth Cellulose Conference, 1988, in press.

5. Graminski, E. L.; Parks, E. J.; Toth, E. E. In Durability of Macromolecular Materials; Eby, R. K., Ed.; ACS Symposium Series No. 95, American Chemical Society: Washington, DC, 1979; pp. 341-355.

6. Segal, L. In Cellulose and cellulose derivatives, Part V; Bikales, N. M. and Segal, L., Eds.; John Wiley & Sons, Inc.: 1971; pp. 719-739.

7. Polyester film encapsulation Library of Congress: Washington, DC, 1980.

8. Shahani, C. J. Abbey Newsletter 1986, 10(2), p.20.

9. Shahani, C. J.; Palmer, C. Unpublished data.

10. Preservation of historical records, National Academy Press: Washington, DC, 1986, p. 25.

11. Mann, J.; Marinnan, H. J. Trans. Faraday Soc. 1956, 52, p. 481.

12. Preston, J. M.; Nimkar, M. V.; Gundavda, S. P. J. Soc. Dyers Col. 1951, 67, p. 169.

13. Sclawy, A. C.   In Preservation of paper and textiles of historic and artistic value II; Williams, J. C., Ed.; American Chemical Society: Wasington, DC, 1982, pp. 217-222.

14. Brickman, W. J.; Dunford, H. B.; Tory, E. M.; Morrison, J. L.; Brown, R. K.   Canadian J. Chem., 1953, 31, pp. 550-563.

15. Rosenthal, A.; Brown, R. K.   Pulp & Paper Mag. Canada, 1950, 48(6), p. 99.

16. Yin, T. P.; Brown, R. K.   Canadian J. Chem., 1959, 37, pp. 444-453.

17. Hermans, P. H.   Physics and chemistry of cellulose fibers, Elsevier Publishing Co.: New York, 1949, p. 192.

18. Shahani, C. J.; Hengemihle, F. H.   In Historic textile and paper materials: Conservation and characterization; Needles, H. L.; Zeronian, S. H., Ed.; American Chemical Society: Washington, DC, 1986, pp. 387-410.

RECEIVED June 9, 1989

Chapter 6

A Reexamination of Paper Yellowing and the Kubelka–Munk Theory

Harald Berndt

Forest Products Laboratory, University of California—Berkeley, 1301 South 46th Street, Richmond, CA 94084

When the rate of yellowing, or brightness reversion of pulps and papers is compared, the twofold interaction of light scattering materials with light needs to be considered. The Kubelka-Munk theory of diffuse reflectance, employing two material constants, the scattering and absorption coefficient, is a useful description of the optical properties of paper. The application of this theory to the study of paper yellowing is reexamined. Changes in scattering coefficients are related to changes in fundamental properties of paper.

The use of optical tests in the characterization of paper materials ranges from color analysis and matching to the standard techniques of chemical analysis by ultraviolet, visible light, and infrared spectroscopy. The test for yellowing, or brightness reversion of pulps and papers is often used as an indicator of their long term stability.

Optical analyses are investigations of the interactions of light with the material being studied. In the case of paper, there are two important types of interactions: impinging photons can be absorbed or they can be reflected at an interface between air and solid material. The absorption characteristics of a material are a function of its chemical composition, which is why their determination is the goal of spectrophotometric analyses. The study of paper yellowing is a special case of spectroscopy, aimed at measuring the rate of formation of material that absorbs light at a specific wavelength, usually near 457 nm.

Transparent materials interact with light only by absorption. This interaction is formulated quantitatively in the Bouguer-Lambert and Beer's Laws (c.f. 1). In paper, however, surface reflection is the dominating type of interaction. This results in very desirable properties like high brightness and opacity, but complicates the interpretation of optical tests with regard to absorption data. The Kubelka-Munk theory attempts to separate the two types of

0097–6156/89/0410–0081$06.00/0

interaction, and this article will reexamine its application to the
study of paper yellowing.

Reflectance spectroscopy

Paper is a highly porous material with many interfaces between air
and the paper fibers. These interfaces between materials of
different refractive indices reflect light by the process known as
regular or Fresnel reflectivity (c.f. 2, 3). Since the interfaces
are orientated at all possible angles to the macroscopic surface, the
resulting reflectance is diffuse, i.e. the intensity of the reflected
light is the same at all angles of observation. The physical
description of diffuse reflectance is known as Lambert's cosine law
(c.f. 2-4). Most of the light impinging on paper will, after
penetrating to some depth and being reflected one or more times at
interfaces, leave the paper toward the side facing the light source.
The fraction of light of a given wavelength exiting to the
illuminated side, expressed as a percentage, is the paper brightness.
Brightness measurement is a special case of reflectance spectroscopy,
the general goal of which is to determine absorption data by
measuring the attenuation of reflected light.

The simplest physical descriptions or theories of light
scattering materials employ two constants, the absorption and
scattering coefficients (k and s). These coefficients are measures
of the changes in the intensity of light passing through a material
due to absorption and reflection, respectively. The best-known and
most successful of these two-parameter theories is that of Kubelka
and Munk (K-M) (5, 6). Several simplifying assumptions regarding the
diffuse reflecting material have to be made in order for a two-
parameter theory to give valid results. Other theories employing up
to eight material constants have been proposed (7). In general, all
theories of diffuse reflection can be simplified or converted to give
the same results as K-M.

In transmission spectroscopy of transparent materials, data on
the light attenuation caused by a known sample thickness suffice to
calculate a function that is linear in the concentration of absorbing
material. To calculate the K-M parameters explicitly, the results of
two independent attenuation measurements need to be known. Commonly,
the reflectances, i.e. the ratios of incident to reflected light
flux, of a layer so thick as to be perfectly opaque (R_∞), and of a
thin layer over a black background (R_0) are determined. The former,
if measured with the proper instruments at a wavelength of 457 nm, is
the TAPPI standard brightness (8). The ratio of R_0 to R_∞, measured
at an effective wavelength of 557 nm, is called the diffuse opacity
of paper (paper backing) (9). Several authors compiled equations
that allow the K-M parameters to be calculated from various other
combinations of attenuation data (6, 10-12). It is possible to
calculate a standard brightness (R_∞) even if a thick, opaque layer of
the sample material is unavailable.

In the original derivation of K-M, the thickness of the
diffusing layer is used as a measure of the optical path length. As
layer thickness has proved inconvenient when working with paper, van
den Akker derived the K-M equations based on weight per unit area, or
basis weight, and showed that all relationships and tabulations of K-

M still hold (13). Following van den Akker, the K-M parameters for pulp and paper are usually expressed in units of area per mass (e.g. cm^2/g). One particularly attractive feature of this approach is the additivity of the specific absorption and scattering coefficients: the coefficients of a composite are the sums of the coefficients of the components, weighted by their mass fraction. This additivity was demonstrated for the absorption coefficient by Giertz (14), who showed that k increases linearly with the mass of dyed fibers added to a pulp. Parsons (15) showed that the scattering coefficient, too, is additive by measuring s of various fiber length fractions of pulps and of their mixtures. The additivity of s holds strictly only when the components are not in appreciable optical contact, but Parsons (15) found that only some fines fractions of chemical pulps deviate from the rule.

The best-known, and probably most important, result of K-M is that the reflectance of a layer of material thick enough to be perfectly opaque is not linearly related to the concentration of an absorbing chemical species, but is given by

$$R_\infty = 1 + k/s - (k^2/s^2 + 2\ k/s)^{\frac{1}{2}} \qquad (1),$$

which can be solved for the ratio k/s, also known as the Kubelka-Munk number, to give

$$k/s = f(R_\infty) = (1 - R_\infty)^2/2R_\infty \qquad (2).$$

This function, sometimes called the remission function (4), is directly proportional to k, and therefore directly proportional to the concentration of an absorbing chemical species. Taking logarithms of Equation 2 yields

$$\log k - \log s = \log f(R_\infty) \qquad (3).$$

Plotting log (k/s) against wavelength or wave number gives an approximation to a real absorption spectrum to the extent that s can be assumed to be constant. Spectra will then be displaced by -log s in the y-direction compared to standard transmission spectra. Plots of log (k/s) are known as characteristic (4) or typical (16) color curves. The scattering coefficient actually decreases slightly with increasing wavelength (15), but the effect on characteristic color curves is negligible.

Some of the problems and limitations of the K-M theory have been discussed by Stenius (17-19). The great sensitivity of the absorption coefficient to small errors in the reflectivity measurement is particularly remarkable. Even though the derivation of K-M assumes perfectly diffuse illumination and reflectivity measurement, the optical geometry of practical measuring apparatus has been a point of discussion (20-24).

Physical interpretation of the Kubelka-Munk parameters

The K-M parameters are purely phenomenological, i.e. they were not derived from fundamental material properties. It is possible,

however, to interpret them in terms of such properties. The
absorption coefficient, k, is directly related to the molecular
absorptivity defined in Beer's law. It has been shown (13) that the
optical path length using diffuse illumination is doubled compared to
illumination with directed, parallel light. Corrections have to be
made for the interactions of absorption and scattering in determining
an effective optical path length (25-27).

The scattering coefficient is more difficult to relate to
fundamental properties. An interpretation of paper reflectance based
upon morphology has been attempted by Scallan and Borch (28, 29).
Their model, viewing paper as composed of idealized, parallel layers,
allows qualitatively correct predictions of changes in scattering
coefficients due to pulp treatments like beating. It has been
argued, however, that this model is contained in K-M, and that
determination of the morphological constants in this multilayer model
uses discretion (30).

A simple and effective rationalization of the scattering
coefficient, s, is given by Robinson (31), who states that

$$s = (\text{const.}) \ (A) \ (R) \qquad (4),$$

where A is the specific surface area per unit mass (cm^2/g), and R is
given by

$$R = (n_r - 1)^2/(n_r + 1)^2 \qquad (5),$$

with

$$n_r = n_1/n_2 \qquad (6).$$

Here, n_1 and n_2 are the refractive indices of the interfacing
materials. The surface reflectivity R, Equation 5, used by Robinson
is the Fresnel reflectivity for normal light incidence (c.f. 4).
Robinson's interpretation explains the well-known linear relationship
between the specific surface area of a paper, as measured by gas
adsorption, and its scattering coefficient (15, 28, 32). It also
rationalizes the observed decrease in scattering and brightness of
wet compared to dry pulp (33, 34).

It is easy to see that R decreases as n_r approaches unity.
Impregnation of a strongly scattering material with a liquid of
similar refractive index can therefore significantly reduce
scattering and make the material amenable for transmission
spectroscopy (35). I found a reduction of scattering power of
Whatman No. 1 filterpaper to 20% of its original value on
impregnation with liquid paraffin. Furthermore, if the scattering
power of a material, impregnated with a series of liquids with
varying refractive index, exhibits a minimum (Figure 1), this minimum
should correspond to the refractive index of the strongly scattering
material.

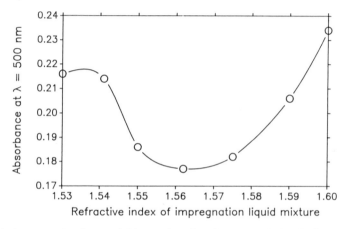

Figure 1. Apparent absorbance of thin wood sections impregnated with liquid mixtures of varying refractive index (1-bromonaphthalene and *n*-butyl-ether) as a function of the refractive index of the liquid.

Factors affecting the scattering coefficient

It can be seen from Equation 4, that the scattering coefficient of a paper changes when the ratio of refractive indices at the interfaces, or the specific surface area of the pulp, are changed. Examples of the effects of changing refractive index ratio, as affected by impregnation of a paper with liquids, has been discussed above. Table I summarizes examples of treatments that change the scattering surface area. It should be noted that not all of the measurable surface area will contribute to light scattering, but only the non-bonded surface area that is not in "optical contact" (15, 36). It is not quite clear how far two surfaces have to be separated in order to have no optical contact (36), but it can be assumed that the necessary separating distance is of the order of magnitude of the wavelength of light. It has been shown that clay coatings scatter light of 457 nm wavelength best when the pore size lies between 500 and 700 nm (37). The observed slight decrease of scattering coefficient with increasing wavelength (c.f. 15, 32) may be attributable to an effective decrease in the surface area which is not in optical contact.

Table I.   Comparing Scattering Coefficients

Cause of Difference in s (pulps compared)	$s_{pulp\ 1}/s_{pulp\ 2}$
- type of pulp[1] (spruce groundwood/bleached jack pine kraft)	2.5
- fines removal[2] (bl. hardwood soda: pulp without fines/whole pulp)	1.1
- pulp fractionation[1] (spruce groundwood: whole pulp/+20 mesh fraction)	2.3
- sheet formation[3] (unbeaten pulp: butanol sheet/water sheet)	1.5
- sheet formation[3] (pulp beaten for 40 min: butanol sheet/water sheet)	4.0
- beating[3] (unbeaten pulp/pulp beaten for 95 min)	1.7
- wet pressing[1] (10 psi/5000 psi)	1.8

[1]: (15); [2]: calculated from data in (15), using additivity of scattering coefficients; [3]: (36).

The decrease in light scattering with increased beating, decreased pulping yield, and increased pressure in wet pressing is attributable to an increase in fiber-to-fiber bonding, which explains the observed inverse relationship between scattering power and pulp strength properties like breaking length (c.f. 38, 39). Casting sheets from butanol suspension results in a significantly increased scattering coefficient because the fiber surfaces are not pulled into optical contact by the capillary forces of the receding water during drying. Fiber bonding and strength properties in such sheets are poor.

The study of paper yellowing

When discussing the durability and permanence of paper, strength properties are of foremost concern. The discoloration of paper with age, referred to as yellowing or brightness reversion, is another important criterion of paper permanence. One of the earliest quantitative investigations of paper yellowing was carried out by Tongren (40), who found that the Kubelka-Munk number, k/s, of a paper increases linearly with the square root of the time of exposure to accelerated aging conditions. Giertz (14) used the difference in k/s before and after heat aging as a measure of pulp discoloration. The said difference, multiplied by 100, was called the post-color number, which has since been used extensively in the study of paper yellowing (c.f. 41).

In Giertz's treatment, as generally in yellowing studies, it is assumed that s remains unchanged during the course of ageing. For chemical changes, this can be shown to be a good approximation, using the additivity of K-M coefficients. I could not find any literature data on structural changes during ageing that affect the scattering coefficient. My own data show a trend towards a decrease in scattering during dry heat ageing. Decreased scattering alone would cause some darkening, but increase in k is clearly the predominant factor. It is possible that the apparent decrease in scattering is due to the observed inverse relationship between k and s (32). Given this uncertainty in the interpretation of changes in s during yellowing, the assumption of constant s seems to be the best working hypothesis, and will be adopted in the following discussion.

Frequently, investigators using the K-M relations have assumed negligible differences in s for differently treated pulps. Tongren (40) found that, with changing rosin content of the paper, the scattering coefficient changed randomly and by small amounts compared to changes in the absorption coefficient. Giertz's (14) definition of the post color number assumes constant s, both during the course of ageing and between pulps being compared. This assumption has been criticized, as changes in interfiber bonding change s (41), and post-color number comparisons favor pulps with higher scattering power (32).

Consider the comparison of pulp ageing rates presented in Figure 2. Bleached rice straw pulps of very low silica content (42, 43) yellowed significantly on dry heat ageing, which is common for nonwood plant fiber pulps (c.f. 42). Extraction with nonpolar solvents, carried out to remove cutin, seemed to reduce the yellowing rate significantly. Plotting k/s against the square root of dry

ageing time resulted in straight lines for the three pulps being
compared (filterpaper was included as a control). All three lines
intersect near a point at k/s = 0. It can easily be seen that such a
family of curves would be produced if only the scattering
coefficients of the pulps were different. The curves' common point
must be the one for k = 0, since the scattering coefficients are
different from each other and must be nonzero. Since real pulps have
an initial k greater than zero, this common point will occur at a
time less than the start of incubation (t = 0). The negative
incubation time is the time that would have been necessary to create
the initial amount of chromophores (k_0) at the given yellowing rate.
In a situation like the one illustrated in Figure 2, the scattering
coefficients must be determined in order to allow unequivocal
interpretation of the data. In this example, the samples with lower
yellowing rate had indeed higher scattering coefficients, but some
real differences remain in the rate of absorption coefficient
increase. Nonpolar solvent extraction of the bleached rice straw
pulp definitely reduces the rate of formation of light absorbing
material during dry heat ageing.

In another example from the literature (44), it can be seen that
ammonia treatment of paper decreased the rate of yellowing
significantly (Figure 3). Following k/s with ageing time leads to
the conclusion that the liquid ammonia treatments are most successful
in reducing yellowing. If the scattering coefficient is determined,
from the published brightness and opacity data, it becomes apparent
that roughly 1/3 of the decrease in yellowing rate from the liquid
ammonia treatments is due to an increase in the scattering power of
the papers.

In both examples discussed here, an increase in scattering
coefficient due to nonaqueous treatments led to an overestimation of
the reduction in yellowing rate affected by these treatments. The
mechanism leading to the increase in scattering coefficient is
probably similar to that discussed above with respect to the
formation of sheets from butanol suspension: in drying from
nonaqueous media, fiber surfaces are not pulled into optical contact
as they would be by the capillary forces of water.

Experimental

Preparation of the rice straw pulps used as examples in Figure 2 is
documented elsewhere (39). My optical measurements on paper were
carried out with a Photovolt Densitometer, Photometer Model 501-A,
Reflection Density Unit # 53, Search Head B, at 445 nm nominal
wavelength. A magnesium carbonate surface was used as a reference
standard of 100 % reflectivity. The scattering power of a paraffin
impregnated filterpaper was calculated from reflectance values over a
black and a white background. Some of the pulp sheets were also
measured using a Hunter Laboratories LABSCAN II, operating at 457 nm.
The values of optical parameters from the two instruments were in
reasonably good agreement.

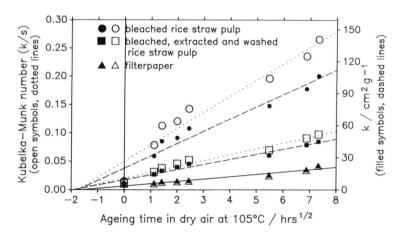

Figure 2: Comparison of yellowing rates of bleached rice straw pulp; bleached rice straw pulp, solvent exchanged and extracted with dichloromethane, and again solvent exchanged and washed with water; and Whatman No. 1 filterpaper (control). The left and right ordinates are scaled so that the two curves for the control coincide (dashed-dotted line).

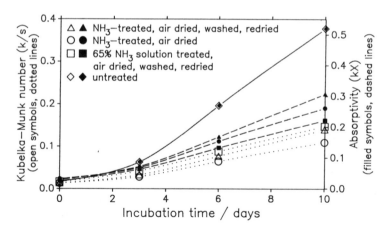

Figure 3: Comparison of yellowing rates of untreated pulp and of pulps subjected to various ammonia treatments (44). The left and right ordinates are scaled so that the curves of the untreated pulp (control) coincide (dashed-dotted curve). (The nomenclature kX for the absorptivity, or absorption power, follows Parson's (15) use)

Summary and conclusions

The Kubelka-Munk theory of diffuse reflectance is a good description of the optical properties of paper. The two parameters of the theory, absorption and scattering coefficient, are purely phenomenological, but are closely related to basic properties of paper. The absorption coefficient is approximately a linear function of the chromophore concentration in the paper. The scattering coefficient is related to the nonbonded fiber surface area in the paper, or the area "not in optical contact," and the Fresnel reflectivity of that surface.

The goal of yellowing studies is the determination of the chromophore formation rate. Therefore, Kubelka-Munk numbers (k/s) should be followed rather than brightness values, since they are linear in chromophore concentration. It is probably safe and advisable to assume that the scattering coefficient of a given pulp is constant during ageing tests. When comparing different pulps, however, their initial scattering coefficients should be calculated explicitly and used to correct the yellowing rates determined in terms of k/s. This will avoid a wrongfully favorable assessment of pulps and papers with high scattering power, or vice versa.

Since the scattering coefficient is a function of the paper structure, and is closely related to some strength properties and to the surface area accessible to chemicals, closer attention to its changes may yield valuable information.

Literature cited

1. Longhurst, R. S. Geometrical and Physical Optics; Longman: London, 1967; Chapter XX.
2. Kortüm, G. Reflectance Spectroscopy; Springer-Verlag: Berlin, 1969; Chapter II.
3. Wendtland, W. M.; Hecht, H. G. Reflectance Spectroscopy; Wiley-Interscience: New York, 1966; Chapter II.
4. Wendtland, W. M.; Hecht, H. G. Reflectance Spectroscopy; Wiley-Interscience: New York, 1966; Chapter III.
5. Kubelka, P.; Munk, F. Zeitschr. f. techn. Phys., 1931, 12, 593.
6. Kubelka, P. J. Opt. Soc. Am., 1948, 38, 448.
7. Duntley, S. Q. J. Opt. Soc. Am., 1942, 32, 61.
8. TAPPI Standards and Suggested Methods; T 452 om-83; Technical Association of the Pulp and Paper Industry, Atlanta, GA.
9. TAPPI Standards and Suggested Methods; T 519 om-80; Technical Association of the Pulp and Paper Industry, Atlanta, GA.
10. Kortüm, G. Reflectance Spectroscopy; Springer-Verlag: Berlin, 1969; Chapter IV.
11. Judd, D. B.; Wyszecki, G. Color in Business, Science and Industry; Wiley: New-York, 1975; 420-438.
12. Robinson, J. V. Tappi, 1975, 58, 152.
13. van den Akker, J. A. Tappi, 1949, 32, 498.
14. Giertz, H. W. Svensk Papperstid., 1945, 48, 317.
15. Parsons, S. R. Paper Trade J., Technical Section, 1942, 115, 314.
16. Kortüm, G. Reflectance Spectroscopy; Springer-Verlag: Berlin, 1969; Chapter V.

17. Stenius, Å. S. Svensk Papperstid., 1951, 54, 663.
18. Stenius, Å. S. Svensk Papperstid., 1951, 54, 701.
19. Stenius, Å. S. Svensk Papperstid., 1953, 56, 607.
20. Stenius, Å. S. J. Opt Soc. Am., 1955, 45, 727.
21. van den Akker, J. A.; Dearth, L. R.; Shillcox, W. M. Tappi, 1963, 46(5), 202A.
22. Stenius, Å. S. Tappi, 1963, 46(11), 183A.
23. van den Akker, J. A.; Dearth, L. R.; Shillcox, W. M. Tappi, 1963, 46(11), 187A.
24. Stenius, Å. S. Tappi, 1965, 48(12), 45A.
25. Butler, W. L. J. Opt. Soc. Am., 1962, 52, 292.
26. Lathrop, A. L. J. Opt. Soc. Am., 1966, 56, 926.
27. Boutelje, J.; Moldenius, S. Svensk Papperstid., 1982, 85, R1.
28. Scallan, A. M.; Borch, J. Tappi, 1972, 55, 583.
29. Scallan, A. M.; Borch, J. Tappi, 1974, 57, 143.
30. Kimura, M.; Iwasaki, Y. Mokuzai Gakkaishi, 1983, 29, 68.
31. Robinson, J. V. Tappi, 1976, 59(2), 77.
32. Hartler, N.; Rennel, J. Papier, 1967, 21, 662.
33. Stenius, Å. S. Tappi, 1969, 52, 942.
34. Hall, F. D. Tappi, 1959, 42(7), 16A.
35. Kuo, M.-L.; Arganbright, D. G.; Zavarin, E. Technical Report No. 35.01.111, 1973, University of California Forest Products Laboratory, Richmond.
36. Ratcliff, F. T. Tappi, 1949, 32, 357.
37. Climpson, N. A.; Taylor, J. H. Tappi, 1976, 59(7), 89.
38. Teder, A. Svensk Papperstid., 1964, 67, 911.
39. Kordsachia, O.; Patt, R.; Rachor, G. Papier, 1988, 42, 261.
40. Tongren, J. C. Paper Trade J., 1938, 107, 76.
41. Spinner, I. H. Tappi, 1962, 45, 495.
42. Berndt, H. M. S. Thesis, Univerity of California, Berkeley, 1982.
43. Brink, D. L.; Merriman, M. M.; Radhakrishna, K.; Berndt, H.; Reddy, M.; Yang, Y. S. Nonwood Plant Fiber Pulping Report No. 18, TAPPI Press, 1988, p. 1.
44. Koura, A.; Krause, Th. Papier, 1980, 34, 555.

RECEIVED June 9, 1989

CONSERVATION AND DEGRADATION OF TEXTILES

Chapter 7

The Stabilization of Silk to Light and Heat

Screening of Stabilizers

M. A. Becker[1], S. P. Hersh, and P. A. Tucker

College of Textiles, North Carolina State University,
Raleigh, NC 27695–8301

A number of stabilizers representing many types have been
screened both singly and in combination to determine
their ability to stabilize silk to light and heat. The
only stabilizer showing any significant improvement in
the first set of experiments was a UV-light absorbing
benzophenone. After 20 days exposure to light, the
stabilized silk fabric retained 22.6% of its original
strength compared with only 13.4% retained by the
unstabilized control. None of these stabilizers improved
the stability to 150°C dry heat. A second set of samples
was then evaluated which included a second UV-light
stabilizer, this time a benzotriazole. In an attempt to
improve the heat stability, the polymeric hindered amine
light stabilizer (HALS) evaluated in the first set was
replaced by two new HALS's. One combination provided
outstanding protection to light. After 20 days, the
strength retained was 86.2% (compared with 31.3% for the
untreated fabric). Unfortunately, the HALS's, which so
dramatically improved the light stability of the treated
silk fabrics, lowered the stability to 150°C heat. Ways
of overcoming this deficiency are being explored.

This paper follows earlier studies (1,2) in which techniques were
developed for studying the degradation of modern silk fabrics
produced by exposing them to dry heat and to radiation from a xenon
arc lamp. Damage was assessed by measuring strength loss, increase
in color, and increases in the ammonia and amino-group contents. The
possibility of determining the mechanism of degradation of
contemporary and historic silk fabrics based on the interrelationship
between the degradation parameters was explored (2,3).
    The objective of this work is to screen stabilizers for their
effectiveness in slowing or preventing the degradation of silk
fabric. The treated fabrics were then subjected to accelerated aging

[1]Current address: Department of Materials Science and Engineering, Johns Hopkins University,
Baltimore, MD 21218

by exposing them to dry heat or to light under a xenon arc lamp. The
extent of degradation of the silk fabrics was then measured by the
procedures listed above.
The stabilizers chosen for evaluation include different types
of heat and light stabilizers selected to represent different
mechanisms of action as well as chemical compositions (4,5). Types
of stabilizers evaluated include: benzotriazole and benzophenone
light stabilizers [ultraviolet (UV) light absorbers], hindered amine
light stabilizers (HALS, catalytic radical scavengers), hindered
phenol heat stabilizers (antioxidant radical scavengers), and
thioester heat stabilizers (antioxidant hydroperoxide decomposers).
In a preliminary trial, one example of a deacidifying agent commonly
used on cellulosic materials was evaluated to determine whether it
would retard silk hydrolysis. Based on reports in the literature
that synergism is possible with certain combinations of light
stabilizers (5), a number of two-component mixtures were also
evaluated. It is not the intent of this study to quantitatively
determine the degradation products or mechanism of silk
deterioration; rather it is to use the relative changes in color,
yarn tensile strength, and nitrogen contents as a means of screening
the stabilizers for their effectiveness in order to select candidates
for further study.

Experimental

Fabric. The silk fabric used in this work was an unweighted plain
woven Chinese silk habutae (Testfabrics, Inc., Middlesex, NJ, style
#605) having 126 ends/in. (37.6 denier), 117 picks/in. (32.6 denier)
and weighing 1.11 oz./yd. The fabric as received had been degummed
[1]. All fabric samples were taken from the same bolt.

Stabilizers. Preliminary set: The stabilizing additives used are
listed in Table I. A preliminary set of fabrics was prepared for
evaluation which consisted of Good-rite 3125 in two concentrations (a
hindered phenol suggested for use on polyamides in B.F. Goodrich Co.
Technical Bulletin GC-60, 1982), Cyanox 1790 (hindered phenol),
Cyanox STDP (thioester), Cyasorb 531 (benzophenone), Chimassorb 944LD
(hindered amine), Wei To #2 (methoxy magnesium methyl carbonate),
Irganox 1098 (hindered phenol), and combinations of Good-rite 3125
with Cyanox STDP or Chimassorb 944LD. Good-rite 3125 and Cyanox 1790
were included in this preliminary set since they had shown some
promise earlier in lowering the generation of amino groups upon
heating (1). The fabrics prepared for this set are listed in Table
II. Although it did not seem to protect silk (1), Wei To #2 was
evaluated since it had proven effective in stabilizing cotton fabric
to degradation by heat (6). Because of their unrealistically high
add-ons, the fabrics in this set were not evaluated further.
Set A: A second set of examples, identified as Set A, was then
prepared for a complete evaluation. As listed in Table III,
stabilizers applied to Set A were Good-rite 2135, Cyanox 1790,
Cyasorb 531, Chimassorb 944LD, Irganox 1098 and Cyanox STDP. Wei To
#2 was eliminated from this set because a precipitate remained on the
surface of the fabric after it was applied in the preliminary set;
the precipitate probably was caused by the unusually high add-on.
The Good-rite 3125 was also combined with the Chimassorb 944LD and
Cyanox STDP to evaluate possible synergistic effects between
combinations of a hindered phenol stabilizer with first a thioester
and then a hindered amine light stabilizer (HALS).

Table I.   Stabilizers Employed, Their Type, Chemical Name and

Supplier

Tradename (type) Source	Chemical Name
Wei To #2 (deacidifier) Wei To Corp.	Methoxy magnesium methyl carbonate
Good-rite 3125 (hindered phenol) B. F. Goodrich Co.	3,5-Di-tert-butyl-4-hydroxyhydro-cinnamic acid triester with 1,3,5-tris(2-hydroxy-ethyl)-s-triazine-2,4,6(1H,3H,5H)-trione
Cyanox 1790 Antioxidant (hindered phenol) American Cyanamide Co.	1,3,5-Tris(4-tert-butyl-3-hydroxy-2,6-dimethylbenzyl)-s-triazine-2,4,6(1H,3H,5H)-trione
Cyasorb UV 531 Light Absorber (benzophenone) American Cyanamide Co.	2-Hydroxy-4-n-octoxybenzophenone
Chimassorb 944 LD (polymeric hindered amine) Ciba-Geigy Corp.	N,N'-bis(2,2,6,6-tetramethyl-4 piperidinyl)-1,6-hexane-diamine polymer with 2,4,6-trichloro-1,3,5-triazine and 2,4,4 trimethyl-1,2-pentanamine
Irganox 1098 (hindered phenol) Ciba-Geigy Corp.	N,N'-Hexamethylene bis(3,5-di-tert-butyl-4-hydroxyhydrocinnamamide
Cyanox STDP Antioxidant (thioester) American Cyanamide Co.	Distearylthiodipropionate
Tinuvin 327 (benzotriazole) Ciba-Geigy Corp.	2-(3',5'-di-tert-butyl-2'-hydroxyphenyl)-5-chlorobenzotriazole
Tinuvin 765 (HALS: hindered amine light stabilizer) Ciba-Geigy Corp.	Bis(1,2,2,6,6-pentamethyl-4-piperidinyl)sebacate(minor component proprietory)
Tinuvin 770 (HALS) Ciba-Geigy Corp.	Bis(2,2,6,6-tetramethyl-4-piperidinyl) sebacate

Table II. Additives Applied to Preliminary Samples Prepared
for Evaluation of Stabilizers on Silk

Stabilizer	Solvent	Add-on Desired(%)	Actual(%)
Good-rite 3125	DMF	1.00	3.50
Good-rite 3125	DMF	3.00	5.40
Cyanox 1790	DMF	3.00	7.70
Cyasorb 531	DMF	3.00	13.80
Chimassorb 944LD	DMF/Chloroform (2:3)	3.00	10.30
Wei To #2	None	3.00	26.50
Good-rite 3125 + Cyanox STDP	DMF/Chloroform (1:2)	1.00 1.00	8.00
Good-rite 3125 + Chimassorb 944LD	DMF/Chloroform (1:2)	1.00 1.00	8.30
Cyanox STDP	DMF/Chloroform (3:11)	3.00	23.70
Irganox 1098	DMF	3.00	8.50

Table III. Stabilizers Applied to Silk Fabrics in Set A

Stabilizer	Solvent	Add-on Desired(%)	Actual(%)
None	DMF	0.0	0.00
Good-rite 3125	DMF	1.0	0.96
Good-rite 3125	DMF	3.0	1.76
Cyanox 1790	DMF	3.0	2.54
Cyasorb 531	DMF	3.0	3.60
Chimassorb 944LD	DMF/Chloroform (1:2)	3.0	2.99
Irganox 1098	DMF	3.0	1.55
Good-rite 3125 + Cyanox STDP	DMF/Chloroform (1:2)	1.0 1.0	2.12
Good-rite 3125 + Chimassorb 944LD	DMF/Chloroform (1:2)	1.0 1.0	2.40
Cyanox STDP	DMF/Chloroform (1:2)	3.0	1.71

Set B: Based on the results of set A discussed below, a third
set, Set B, was prepared (Table IV). Cyasorb 531 (a benzophenone UV
light absorber) was found to be the most effective stabilizer in Set
A. It was therefore rerun at an even lower add-on in Set B, and
since it was the only UV light absorber included in Set A, a second
type of UV absorber, Tinuvin 327 (a benzotriazole) was included in
Set B. Two new non-polymeric hindered amine light stabilizers (HALS)
Tinuvin 770 and Tinuvin 765 were also evaluated in Set B with the
hope that they might be more effective than the polymeric HALS
(Chimassorb 944LD) included in Set A which was quite detrimental to
the heat stability. Combinations of these four additives were also
applied in Set B. The three new compounds evaluated have been
reported to be effective light stabilizers for polyester and
polyamide fibers (Capocci, G., Ciba Geigy Corporation, Ardsley, NY,
personal communication, 1986).

Table IV. Stabilizers Applied to Silk Fabrics in Set B

		Add-on	
		Desired	Actual
Stabilize	Solvent	(%)	(%)
None	None	0.0	0.0
None	DMF	0.0	0.0
Cyasorb 531	DMF	2.0	1.4
Tinuvin 327	DMF/Xylene (20:3.5)	2.0	2.0
Tinuvin 765	DMF	2.0	2.6
Tinuvin 770	DMF	2.0	1.2
Cyasorb 531 + Tinuvin 765	DMF	2.0 2.0	6.6
Cyasorb 531 + Tinuvin 770	DMF	2.0 2.0	3.8
Tinuvin 327 + Tinuvin 765	DMF/Xylene (20:3.5)	2.0 2.0	4.0
Tinuvin 327 + Tinuvin 770	DMF/Xylene (20:3)	2.0	3.1

Application of Stabilizers. The fabrics were prepared by applying
stabilizers dissolved in solvents selected for their ability to
dissolve the additive as well as to swell silk. The concentration of
additive in solution was adjusted to give the desired fabric add-on
assuming the fabrics would have 100% wet pick-up. If the fabric wet
pick-up is indeed 100%, the dried fabric would have the desired add-
on.
        N,N-Dimethyl Formamide (DMF) was the preferred solvent for all
applications because of its excellent swelling characteristics (7).

However, for some stabilizers it was necessary to add chloroform or
xylene to the DMF to dissolve the additive. In the preliminary set,
fabric samples 15 in. by 15 in. were soaked in the stabilizer
solutions for 30 minutes and then run through a Bronco Model 110
padder at a pressure of 100 kPa in an attempt to obtain the
appropriate add-on. The padded fabrics were then set on pin-frames
to air dry. As shown in Table II, however, the add-ons were all
unacceptably high. It is believed that this procedure, rather than
squeezing excess solution from the very light-weight silk fabrics,
allowed the solvent to evaporate while the sample was running through
the padder, thus leaving a precipitate on the fabric surface. For
this reason Sets A and B were prepared without padding.
     Instead, for Set A, 18 cm by 18 cm. fabric samples were soaked
in the stabilizer solution for 30 minutes and then immediately placed
on pin-frames to drain and dry. As shown in Table III, the add-ons
obtained by this procedure were much closer to those expected.
     Set B was prepared in substantially the same way, but with
minor alterations. Larger samples were used, 15 cm by 91 cm. The
samples were loosely folded and allowed to soak in the stabilizer
solution for 30 minutes, then rinsed with distilled/deionized water
to remove any stabilizer left on the surface. The fabrics were then
hung over a glass rod and allowed to air dry. Again, the add-ons
were much closer to those desired (Table IV).
     Because the add-ons of the preliminary set of samples were so
unrealistically high, these samples were not evaluated further. The
deviations from the desired add-ons for both Sets A and B might be
due to the differing substantivities of the stabilizers.

Accelerated Aging. Thermal: The procedure developed earlier for
thermal aging was followed (1). Fabric pieces measuring 15 cm by 15
cm were placed in a forced convection oven preheated to 150°C on
racks covered with a Fiberglas screen (7 cm x 3 cm mesh). The screen
was used to prevent any enhanced degradation that might occur from
direct contact with the metal rack. Each treated fabric was heated
for up to four days in increments of one day. After heating, the
fabrics were immediately placed in a desiccator containing silica gel
to keep the samples dry while cooling.
     Light: For aging by exposure to light, 20 cm x 7 cm pieces of
fabric were mounted in standard specimen holders as specified in
AATCC test method 16E-1982, "Colorfastness to Light: Water-Cooled
Xenon Arc Lamp, Continuous Light" (8). The fabrics were exposed to
light for up to twenty days in two-day increments in a water-cooled
xenon arc Weather-ometer, model ES 25 (Atlas Electric Devices,
Chicago, IL). The Weather-ometer was operated at 50°C with an arc
intensity of 1500 watts. Relative humidity in the sample chamber was
maintained at 30±5% RH. Each sample received 108 kJ/m^2/nm of energy
passing through a narrow band path filter (420 nm) for each 2-day
interval. The samples were immediately placed in acid-free tissue
paper upon removal from the Weather-ometer. Unfortunately, a direct
comparison cannot be made between Sets A and B exposed in the
Weather-ometer because the air supply to the Weather-ometer became
partially clogged sometime while Set B was being exposed. As a
result, the RH could not be maintained at 30%, but probably remained
below this value during much of the exposure. In general, the lower
the humidity during exposure to light, the less the degradation (9).
Therefore only samples exposed in the Weather-ometer at the same time
can be compared with each other.

Parameters to Measure Degradation. Breaking Strength: Warp yarns
were extracted from the fabrics and their breaking loads were
determined at a gauge length of 5.0 cm and a rate of extension of 50
mm/min on a tensile testing machine (Instron Model 1123, Instron
Corporation, Canton, MA) as set forth in ASTM Test Method D2256-80,
"Breaking Load (Strength) and Elongation of Yarn by Single-Strand
Method" (10). Normally 21 measurements were made from each fabric
sample. The fabrics were allowed to equilibrate under standard test
conditions (21±2°C, 65±2% RH). The breaking strength of the yarns
extracted from the control fabric was 100.0 gf (2.6 gf/denier) with a
standard deviation of approximately 8 gf. The coefficient of
variation of breaking strength of the degraded samples was
approximately 10%.

Color Change: It is well known that silk discolors on exposure
to heat and light. The color change is therefore taken as one
measure of the extent of degradation. The color of each treated
fabric sample was measured on a Diano Match-Scan spectrophotometer
against a standard untreated silk fabric sample mounted on a white
tile background. Because of the small sample size available, only
one layer of the fabric was evaluated. This limitation introduces
some variation within each sample measurement (11). The color
differences (ΔE_{ab}^*) are reported in CIEL*a*b* Color Difference Units
(CDU) for Illuminant D_{65} calculated by the equation (2,12):

$$\Delta E_{ab}^* = [(\Delta L^*) + (\Delta a^*) + (\Delta b^*)]^{1/2}$$

where ΔL^* is the change in lightness, from lighter (+) to darker (-),
Δa^* is the change in shade from red (+) to green (-), and Δb^* is the
change in shade from yellow (+) to blue (-) with respect to the
standard (untreated white silk fabric). Three measurements were
averaged for each fabric.

Amino-Nitrogen Content: Each fabric to be analyzed was first
ground in a Wiley mill fitted with a No. 40 mesh screen. The
concentration of amino groups was then determined colormetrically by
the reaction of 20 mg of the ground silk fabric with ninhydrin using
the procedure described earlier (1,2,13). The reaction of ninhydrin
with amino acids, as well as ammonia and some other amines, forms a
compound known as Ruhmann's purple which has a maximum absorption at
570 nm (14). For the screening experiments made here in which the
major interest is assessing the extent of degradation, the results of
this analysis will be reported simply as the total concentration of
ninhydrin reaction products measured expressed as μmol of amino
groups per gram of silk. The calibration was based on the å-amino
acid d,1-leucine (2).

Ammonia Content: The concentration of ammonia nitrogen present
in the water-extractable components of the fabric was determined
colormetrically by Nessler's reaction (2,15). Nessler's Reagent
reacts with ammonia released from ammonium salts by KOH producing a
yellow compound under alkaline conditions by the following reaction:

$$2K_2HgI_4 + NH_3 + 3KOH \rightarrow Hg_2OINH_2 + 7KI + 2H_2O$$

A 50 mg sample of each ground fabric was introduced into a 50
mL Erlenmeyer flask, and 20-30 mL of deionized/distilled water was
added. After approximately 15 minutes, each solution was filtered
through ashless filter paper into a 100 mL volumetric flask. A 2 mL
aliquot of Nessler's Reagent (APHA, Fisher Scientific Company) was

added to the 100 mL solution. After at least 10 minutes, but not
longer than 20 minutes, the absorbance of the solution at 425 nm was
measured on a Bausch and Lomb Spectronic 20 spectrophotometer.

Results and Discussion

Although the treated samples were heated for up to four days in
increments of one day and exposed to light for up to 20 days in 2-day
intervals, not all the degraded fabrics were evaluated. The heat-
degraded samples in Set A were measured after two and four days
exposure while those in Set B were measured after daily exposure.
The light degraded samples in Set A were measured after four-day
exposure intervals and those in Set B were measured either after 2-
day intervals (tensile strength and color) or 4-day intervals
(ammonia and amino-group content). Since the results observed and
conclusions reached on properties measured during intermittent
exposures are similar to those based on the complete 4-day heating
and 20-day light exposures, only the results for the longest
exposures will be reported and discussed here. For those interested,
the remaining data are available elsewhere (4).

Strength Loss. The strengths of the artificially aged samples are
reported in Table V. After 20-day light exposure, the only sample in

Table V. Yarn Tensile Strength Retained After Artificial Aging of
Fabrics in Sets A and B. (Four Days at 150°C
or 20 Days of Light Exposure)

SET A			SET B		
Stabilizer	Heat (%)	Light (%)	Stabilizer	Heat (%)	Light (%)
DMF	37.9	14.8	None	31.4	33.3
Good-rite 3125 (0.96%)	33.6	16.4	DMF	19.7	28.5
Good-rite 3125 (1.76%)	33.5	16.4	Cyasorb 531	27.5	38.2
Cyanox 1790	28.7	15.3	Tinuvin 327	24.6	27.9
Cyasorb 531	30.2	24.9	Tinuvin 765	5.1	71.0
Chimassorb 944LD	19.4	16.1	Tinuvin 770	5.0	48.4
Irganox 1098	31.4	15.2	Cyasorb 531 + Tinuvin 765	4.7	92.1
Good-rite 3125 + Cyanox STDP	26.5	11.9	Cyasorb 531 + Tinuvin 770	5.1	62.5
Good-rite 3125 + Chimassorb 944LD	29.6	15.7	Tinuvin 327 + Tinuvin 765	4.7	54.3
Cyanox STDP	33.0	12.3	Tinuvin 327 + Tinuvin 770	4.2	60.4

Set A that showed any significant improvement over DMF-treated
control (14.8% strength retention based on the tensile strength of
the original yarn which was 100 gf) was Cyasorb 531 (24.9 gf), a UV
light-absorbing benzophenone. The strength retained by the fabric
treated with the mixture of Good-rite 3125/Cyanox STDP (2.12% add-on)
was 11.9 %, even less than that of the DMF sample. None of the heat-
aged samples in Set A retained their strength as well as did the DMF
sample: all showed substantially lower values after four days of
heating at 150°C.
     The strength retained by a number of samples in Set B after
exposure to light showed considerable promise. The untreated and DMF
silk samples retained 33.3% and 28.5% of their initial strengths
after 20 days of light exposure respectively. In contrast, after the
same 20-day light exposure, the Cyasorb 531/Tinuvin 765, a
combination of the benzophenone examined in Set A and a HALS,
provided the best protection with a strength retention (S.R.) of
92.1%. The strength retentions of the samples treated with Tinuvin
765 along and with Cyasorb 531 alone were 71.0% and 38.2%,
respectively. Thus some substantial synergism is evident by using
these two stabilizers together. The Cyasorb 531 combined with
Tinuvin 770, the second HALS introduced in Set B, also had a good
strength retention, 62.5%, again showing synergism with the Cyasorb
531 since the Tinuvin 770 used alone retained only 48.4% strength.
In contrast, the second UV-sorber, this time a benzotriazole, Tinuvin
327 introduced into Set B, was not effective when used alone (27.9%
S.R., even less than the control), and it even reduced the
effectiveness of one of the two HALS used (Tinuvin 765, from 71% S.R.
when used alone to 54.3%).
     Unfortunately, none of the treatments in Set B showed any
promise in stabilizing silk to thermal aging. In fact, all the
samples that were effective in stabilizing silk to light were very
detrimental to heat stability.
     Although most studies of accelerated heat aging indicate that
the Arrhenius relationship applies to cotton and other cellulosic
materials such as paper (16, 17, 18, 19), it is possible that the
thermal aging conditions used here for silk are too severe, thus
possibly causing some additional mechanisms to come into play that do
not occur during natural aging. Additional studies utilizing other
thermal aging conditions would have to be made to determine whether
these stabilizers might indeed be useful in stabilizing silk fabric
to thermal aging under ambient or moderately elevated conditions as
suggested by de la Rie (5).

Color Change. The increases in color of the silk fabrics treated
with stabilizers after artificially aging are shown in Table VI.
Cyasorb 531, the UV-light absorbing benzophenone stabilizer, was the
only sample in Set A which discolored less (4.7 CDU) than the DMF-
treated control silk sample (5.5 CDU) after a 20-day exposure to
light (a decrease of 14.5%). After heating for 4 days at 150°C, all
of the samples in Set A except Good-rite 3125 (0.96% add-on)
increased in color. The stabilizers in Set B which showed such
striking improvements in strength retention after 20-day light
exposure also demonstrated dramatic improvements in color
stabilization. The combination of Cyasorb 531/Tinuvin 765 discolored
only one-half as much (2.9 CDU) as did untreated silk (5.9 CDU). The
performance of the Cyasorb 531/Tinuvin 770 and Tinuvin 765 alone were
almost as striking - they discolored 4.0 and 4.1 CDU, 32.1% and

30.5%, respectively, less than did the control. Unfortunately, these samples discolored more upon heat aging than did the controls: 53.6%, 50.4%, and 56.8% after 4 days respectively, for the Cyasorb 531/Tinuvin 765 (55.8 CDU), Cyasorb 531/Tinuvin 770 (54.7 CDU), and Tinuvin 765 (57.0 CDU) over the untreated silk sample (36.3 CDU).

Table VI.   Total Color Change ΔE^{*}_{ab} (CIEL*a*b* Color Difference Units, CDU) After Aging of Sets A and B. (4 Days at 150°C or 20 Days of Light Exposure)

SET A			SET B		
Stabilizer	Heat (CDU)	Light (CDU)	Stabilizer	Heat (CDU)	Light (CDU)
DMF	35.1	5.5	None	36.3	5.9
Good-rite 3125 (0.96%)	35.1	5.7	DMF	40.7	5.8
Good-rite 3125 (1.76%)	35.9	5.7	Cyasorb 531	37.4	5.4
Cyanox 1790	39.4	6.4	Tinuvin 327	38.9	5.8
Cyasorb 531	37.8	4.7	Tinuvin 765	57.0	4.1
Chimassorb 944LD	40.2	6.4	Tinuvin 770	53.5	5.0
Irganox 1098	41.8	5.8	Cyasorb 531 + Tinuvin 765	55.8	3.0
Good-rite 3125 + Cyanox STDP	39.6	6.3	Cyasorb 531 + Tinuvin 770	54.7	4.0
Good-rite 3125 + Chimassorb 944LD	43.9	6.2	Tinuvin 327 + Tinuvin 765	51.7	5.1
Cyanox STDP	36.8	6.2	Tinuvin 327 + Tinuvin 770	53.9	4.9

Ammonia Concentration. The ammonia concentrations of the stabilized fabrics after aging are listed in Table VII. All the stabilizers in Set A performed better than the DMF sample after heating 4 days at 150°C. The best protection was provided by the four samples containing Good-rite 3125 (ammonia concentrations varying from 54.7 to 64.5 μmol/g silk and the STDP sample (61.2 μmol/g silk). After 8 days of artificial aging by light, however, the best protection was obtained with Cyasorb 531, the UV sorber, 18.7 μmol/g silk. The samples exposed to light 12 days or longer developed some turbidity during the analysis (2,4). Hence, comparisons of ammonia contents of the samples were made after 8 days exposure to light.

Although none of the treated samples in Set B showed any improvement over the untreated silk after heating 4 days at 150°C, Cyasorb 531 again was the best of the lot (65.8 μmol/g silk). The best sample in Set B after artificial light aging for 8 days was Tinuvin 765, one of the HALS (14.2 μmol/g silk, a 31.4% reduction

over that of the control). Cyasorb 531 and Tinuvin 770 also provided
good protection.

Table VII. Ammonia Content (μmol/g silk) After Artificial Aging of
Fabrics in Sets A and B.(4 Days at 150°C or 8 Days
Light Exposure, Control Silk Fabric = 9.6)

| SET A | | | SET B | | |
Stabilizer	Heat	Light	Stabilizer	Heat	Light
DMF	99.8	29.2	None	63.8	20.7
Good-rite 3125 (0.96%)	59.9	27.3	DMF	90.6	20.1
Good-rite 3125 (1.76%)	61.2	29.9	Cyasorb 531	65.8	19.4
Cyanox 1790	91.3	64.5	Tinuvin 327	71.7	25.9
Cyasorb 531	86.7	18.7	Tinuvin 765	98.5	14.2
Chimassorb 944LD	92.6	29.2	Tinuvin 770	112.8	19.4
Irganox 1098	72.3	40.0	Cyasorb 531 + Tinuvin 765	99.1	18.1
Good-rite 3125 + Cyanox STDP	54.7	32.5	Cyasorb 531 + Tinuvin 770	100.4	16.2
Good-rite 3125 + Chimassorb 944LD	64.5	27.3	Tinuvin 327 + Tinuvin 765	108.3	19.4
Cyanox STDP	61.2	33.8	Tinuvin 327 + Tinuvin 770	105.0	20.7

Amino Nitrogen Content: Ninhydrin Method. The concentration of the
Ninhydrin Reaction Products (NRP) (expressed in terms of the amino-
group concentration in μmol/g silk) of the samples evaluated are
listed in Table VIII. In Set A, all of the samples, except Cyanox
1790 and Cyanox STDP generated somewhat less NRP after 4 days of
heating at 150°C than did the DMF sample. The best sample (Good-rite
3125/Chimassorb 944LD), however, generated only 11.1% less NRP than
did the DMF-treated sample. All of the stabilized samples generated
more NRP after 20 days of light exposure than did the DMF sample.
The best performer was again the UV-sorber Cyasorb 531 followed
closely by Good-rite 3125 (1.76% add-on) which generated increases of
only 6.6% and 5.3%, respectively, over DMF.
      In Set B, after 4 days of heat aging at 150°C, all the
stabilizer-treated fabrics had higher concentrations of NRP than did
the untreated silk sample. In contrast, after 20 days of light
exposure, the same three treatments which have shown the greatest
improvements in strength retention, color, and ammonia concentration
again gave the best performance over the others: Cyasorb 531/Tinuvin
765 (61.3 μmol/g silk), 46.4% reduction over the untreated silk
sample (114.4 μmol/g silk), followed by a 36% reduction for both

Cyasorb 531/Tinuvin 770 and Tinuvin 765 (73.3 μmol/g silk). The DMF
(120.3 μmol/g silk) and Tinuvin 327 (116.7 μmol/g silk) samples
generated higher concentrations of NRP than did the untreated silk
after 20 days of light exposure.

Table VIII. Concentration (μmol/g silk) of Ninhydrin Reaction
Products After Artificial Aging of Fabrics in
Sets A and B. (4 Days at 150°C or 20 Days of Light
Exposure, Control Silk Fabric = 55.4)

SET A			SET B		
Stabilizer	Heat	Light	Stabilizer	Heat	Light
DMF	92.4	91.9	None	91.2	114.4
Good-rite 3125 (0.96%)	85.0	111.9	DMF	112.4	120.3
Good-rite 3125 (1.76%)	89.4	96.8	Cyasorb 531	93.2	97.8
Cyanox 1790	92.4	99.1	Tinuvin 327	99.8	116.7
Cyasorb 531	89.4	95.8	Tinuvin 765	148.1	73.3
Chimassorb 944LD	90.4	117.5	Tinuvin 770	125.1	103.7
Irganox 1098	91.9	120.3	Cyasorb 531 + Tinuvin 765	155.3	61.3
Good-rite 3125 + Cyanox STDP	85.0	114.4	Cyasorb 531 + Tinuvin 770	137.9	73.3
Good-rite 3125 + Chimassorb 944LD	82.2	107.8	Tinuvin 327 + Tinuvin 765	136.1	86.8
Cyanox STDP	93.7	109.8	Tinuvin 327 + Tinuvin 770	143.0	84.0

Conclusions

One of the treatments examined showed exceptional ability to
stabilize silk to light. The system consisted of a mixture of
Cyasorb 531 (a UV-light absorbing benzophenone) and Tinuvin 765 (a
hindered amine light stabilizer or "HALS"). All the properties
measured showed substantial improvement over unprotected silk. For
example, after 20 days exposure to a Xenon arc light, 92.1% of the
breaking strength was retained (compared with 33.3% for the
unstabilized silk). At the same time, the color of the stabilized
sample increased only 3.0 units compared with nearly twice that for
the control (5.9 units). The combination of the two stabilizers was
considerably better than either stabilizer used alone. Two other
mixtures, both containing an ultraviolet absorber stabilizer and a
HALS, improved the light stability of the silk substantially as did
one of the HALS used alone.

Unfortunately, none of the stabilizers screened improved the heat stability, and those containing any of the HALS used in the study degraded the heat stability very badly. The heat aging was carried out at 150°C, a temperature selected to reduce the tensile strength of the silk to about 20% after a reasonable heating time (1). Studies have shown that the accelerated thermal aging of cotton at temp-eratures up to 150°C (16) and 190°C (17) follow the Arrhenius relationship, i.e., the degradation rate measured at a high temperature, for example, can be used to predict the degradation rate at a lower temperature. However, it has not been demonstrated that the heat degradation of silk follows the Arrhenius equation up to the 150°C used here as an accelerated heat aging test. In addition, it has been suggested that HALS, the class of stabilizers that so drastically reduced the strength and increased the discoloration of silk after heating at 150°C, may perform better during natural aging than is predicted from accelerated aging (20). Thus it is evident that the effects of thermal aging should be examined at lower temperatures to determine whether 150°C is unrealistically high for evaluating thermal stability, i.e., if the degradation reactions and mechanisms, especially those associated with the presence of HALS, might not be as severe at lower temperatures. A second approach would be to seek less basic HALS, which would be less detrimental to the stability of silk.

Both of these approaches are being explored in an effort to take advantage of the striking improvements in light stability achieved without incurring the real, or at least implied, problems of poor heat stability. Additional stabilizers will continue to be screened and optimum formulations will be selected.

Acknowledgements

This investigation was supported in part by the National Museum Act which was administered by the Smithsonian Institution. We are grateful for this assistance.

Literature Cited

1. Kurupillai, R. V.; Hersh, S. P.; Tucker, P. A. In Historic Textile and Paper Materials: Conservation and Characterization; Williams, J. C., Ed.; Advances in Chemistry Series No. 212; American Chemical Society: Washington, DC, 1986; pp. 111-27.

2. Hersh, S. P.; Tucker, P. A.; Becker, M. A. In "Archaeological Chemistry IV"; Allen, R., Ed.; Advances in Chemistry Series No. 220; American Chemical Society: Washington, DC, 1989, pp. 429-450.

3. Becker, M. A.; Hersh, S. P.; Tucker, P. A.; Waltner, A. W. Prepr. ICOM Comm. Conserv. 8th Triennial Meet., Grimstad, K., Ed.; The Getty Conservation Institute: Los Angeles, CA, 1987; pp. 339-44.

4. Becker, M. A. "The Stabilization of Silk to Light and Heat." M.S. Thesis, North Carolina State University, Raleigh, NC, 1988.

5. de la Rie, E. R. Studies in Conservation, 1988, 33, 9-22.

6. Kerr, N.; Hersh, S. P.; Tucker, P. A. Prepr. ICOM Comm. Conserv. 7th Triennial Meet. 1984, Copenhagen, Denmark, 84.9.25-84.9.29.

7. Shimizu, Y. J. Seri. Sci. Japan 1983, 47, 417-420.

8. AATCC Technical Manual; American Association of Textile Chemists and Colorists: Research Triangle Park, North Carolina, 1982, Vol. 57, pp. 160-62.

9. Egerton, G. S. Text. Res. J. 1948, 18, 659-69.

10. Annu. Book ASTM Stand. 1983, 07-01, 401-10.

11. Stearns, E. I.; Prescott, W. B. In Color Technology in the Textile Industry; Celikiz, G. and Kuehni, R., Eds.; American Association of Textile Chemists and Colorists: Research Triangle Park, NC, 1983; pp. 63-66.

12. Billmeyer, Jr., F. S.; Saltzman, M. Principles of Color Technology, 2nd Ed.; John Wiley & Sons, New York, 1981; pp. 63, 103.

13. Knott, J.; Grandmaire, M.; Thelen, J. J. Text. Inst. 1981, 72, 19-25.

14. Moore, S.; Stein, W. H. J. Bio. Chem. 1956, 176, 367-88.

15. Standard Methods for the Examination of Water and Wastewater; 14th ed., APHA, Washington, DC, 1976; pp. 407-415.

16. Block, I.; Kim, H. K. In Historic Textile and Paper Materials; Needles, H. L.; Zeronian, S. H., Eds.; Advances in Chemistry Series No. 212, American Chemical Society, Washington, DC, 1986; pp. 411-25.

17. Cardamone, J. M.; Brown, P. In Historic Textile and Paper Materials; Needles, H. L.; Zeronian, S. H., Eds.; Advances in Chemistry Series No. 212, American Chemical Society, Washington, DC, 1986; pp. 41-75.

18. Baer, N. S.; Indicator, N. In Preservation of Paper and Textiles of Historic and Artistic Value; Williams, J. C., Ed.; Advances in Chemistry Series No. 164, American Chemical Society, Washington, DC, 1977; pp. 286-313.

19. Gray, G. G. In Preservation of Paper and Textiles of Historic and Artistic Value; Williams, J. C., Ed.; Advances in Chemistry Series No. 164, American Chemical Society, Washington, DC, 1977; pp. 336-351.

20. Schirmann, P. J.; Dexter, M. In Handbook of Coating Additives; Calbo, L. J., Ed.; Marcel Dekker, New York, 1987; pp. 225-70.

RECEIVED February 22, 1989

Chapter 8

The Conservation of Silk with Parylene-C

Eric F. Hansen and William S. Ginell

Getty Conservation Institute, 4503B Glencoe Avenue,
Marina del Rey, CA 90292

Parylene-C is a vapor-deposited, colorless,
transparent polymer that may have applications
in the consolidation of fragile, porous or
fibrous art objects. Tests were conducted to
determine the effects of a 0.75 μm Parylene-C
coating on the tensile and appearance
properties of modern and historic silk fabrics.
Accelerated thermal and light aging of coated
and uncoated silk were conducted and
degradation rates were determined for modern
silk fabric. Thermal degradation rates of the
tensile properties (breaking-load, strain-to-
break and energy-to-break) and the rate of
yellowing of a modern silk fabric were
unaffected by the presence of the coating.
Photo-degradation of coated silk fabric
resulted in a reduction of tensile properties
and yellowing of the coating that was exposed
to xenon-arc radiation filtered to simulate
outdoor exposure to sunlight. No color
developed in free films of Parylene-C after
exposure to the source that was filtered
through a 400 nm cut-off filter. The effect of
coating structurally strong, modern silk was an
increase in all tensile properties with the
exception of the initial modulus. Coating
comparatively weak, historic silk increased the
breaking-load, energy-to-break and the initial
modulus but did not result in an increased
strain-to-break. This method of consolidation
could be considered for the conservation of
very fragile silk where added strength is the
primary consideration and exposure to
ultraviolet light is minimized.

0097–6156/89/0410–0108$07.50/0
© 1989 American Chemical Society

Historic silk artifacts are often found to be extremely fragile owing either to exposure to certain environmental conditions (1) or to harmful processing procedures such as dying or weighting with heavy metal salts (2). A conservation technique frequently used for the consolidation of crumbling silk fabrics is to use an adhesive to attach the silk to a support backing (3). The choice of an appropriate adhesive presents difficulties because of problems with yellowing, development of brittleness, and biodeterioration that may occur with time (4).

An alternative treatment for weak and fragile silk, one that has not been explored previously, is consolidation using a conformal, polymeric coating. Conformal coatings penetrate porous materials and coat all accessible surfaces with a thin, uniform film. In a woven fabric, adjacent or intersecting threads are reinforced and bonded by the film. In this paper, an application of Parylene-C is considered for the consolidation of highly degraded silk fabrics. Parylene-C is a conformal coating widely used in industrial, space, electronics and medical applications(5).

When considering the use of new materials for the conservation of objects of cultural or historic significance, several important factors must be addressed: will the treatment degrade the physical, chemical or appearance properties of the object; will the treatment enhance these properties; and, if so, will the enhanced properties remain stable long enough so that the "useful lifetime" of the object will be substantially prolonged? This paper provides quantitative information that a conservator can use in weighing the effects of the treatment on degraded silk against other possible treatments or no treatment at all. Accelerated aging techniques were used to model the change in tensile and appearance properties of silk with time and under defined exposure conditions. The results of the aging experiments can then be used by the conservator in defining a "useful lifetime" for Parylene-C coated silk. Consideration must be given, in assessing the durability of the composite, to the differing environmental situations that would be encountered in the display or storage of silk artifacts.

Parylene-C, the trade name of the film formed from the Union Carbide Corp. brand of dichloro-p-xylylene, is a vapor deposited film formed by the reaction shown in Figure 1. The solid dimer(I) is vaporized and pyrolyzed at 650°C to 750°C to the reactive olefinic monomer, chloro-p-xylylene(II), which polymerizes on cool surfaces in the low pressure deposition chamber to form the crystalline linear polymer poly(chloro-p-xylylene)(III) (5). The transparent, colorless and pinhole-free film has a high tensile strength and elongation; a low permeability to water and oxygen; a glass transition temperature of 95°C; and, is insoluble in aqueous and organic solvents (6).

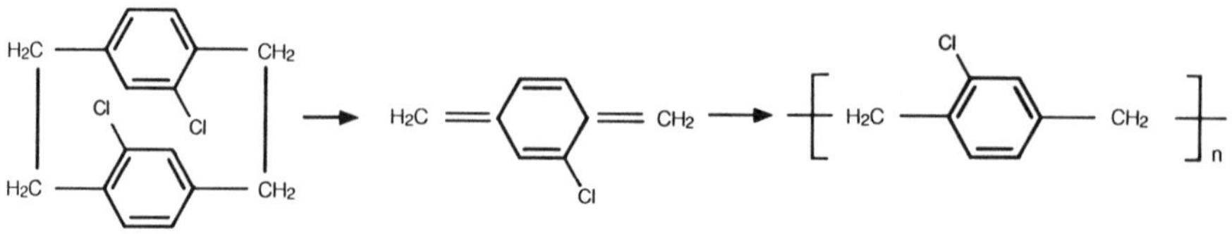

(I)   (II)   (III)

Figure 1. Polymerization of Parylene-C.

Coating thickness can be controlled by varying the initial
amount of dimer used. Thicknesses between 0.5 and 25 μm are
most commonly deposited.

Parylene-C is of potential interest in conservation
because of the unique application method that can result in
the formation of thin, uniform, and strong films within
porous or fibrous materials. However, the polymer has the
disadvantages of being a non-reversible consolidant that is
sensitive to ultraviolet-induced oxidation.

The effects of a Parylene-C coating on silk were
investigated by measuring the tensile properties and
determining the color of coated and uncoated silk fabric.
The color was also determined for free films of Parylene-C.
Accelerated thermal and photolytic aging tests were carried
out to determine either if the coating adversely affected
the normal property degradation rates of modern silk or if
the coating, acting as a barrier to water vapor or light,
reduced the silk degradation rate. For structurally weak
and fragile historic silk, it was important to determine if
a Parylene-C coating would decrease fragility sufficiently
to allow handling, mounting and display of the artifact.

Experimental

Materials. The modern silk fabric tested was Testfabrics
style 607 broadcloth, degummed but undyed and unbleached.
The fabric yarn was 140/2 cotton count and the warp and
filling threads/0.01 m are 40 and 33, respectively. Fabric
weight was approximately 56 gm/m².

The fabric samples tested were 0.17 m x 0.25 m and were
coated with a 0.75 μm layer of Parylene-C. Similarly, a
series of fabric samples was obtained with coating
thicknesses varying from 0.25 to 1.5 μm (the coating was
performed at Nova Tran Corporation, Clear Lake, Wisconsin).

Two types of historic fabric were tested: a blue silk
tabby-woven (plain woven) fabric, or silk taffeta, weighted
with Sn and Fe, which was originally a Hollywood movie
costume silk lining from the 1920's; and, a cream-colored,
unweighted silk fabric with a self-patterned band of tabby,
weft-faced twill and warp-faced satin (a selvedge exists on
one side of the fabric making clear that the patterning
runs in the direction of the weft) that had a mid-19th
century provenance. Samples of each material were coated
with both 1 and 5 μm layers of Parylene-C. In addition to
the coated fabrics, free Parylene-C films, 12 μm thick,
were also studied.

Thermal Exposures. The thermal aging tests on coated and
uncoated samples were conducted in Precision forced
convection ovens. Temperature uniformity was ±1 C, which
was determined with a thermocouple array that was mounted
in the oven containing a set of dummy samples. No change
in temperature uniformity was observed as a result of
removing samples periodically for analysis. Test
temperatures were 150 °C, 110 °C, 90 °C, 80 °C and 70 °C for

maximum exposure times of 7, 45, 118, 109 and 158 days,
respectively. Samples were removed for analysis at
intervals during each constant temperature exposure.
For experiments involving humidity variation, a Hot
Pack Temperature-Humidity Chamber, Model # 435314, with
digital humidity and temperature control, was used.
Humidity levels were maintained at <5%, 50%, 70% or 90% ±2%
RH and a constant temperature of 90°C for a fixed time of
17 days. Samples of historic silk fabric were exposed at
90°C and 80% RH for either 10 or 30 days to induce further
deterioration of mechanical properties. These materials
were then coated with 1 and 5 μm of Parylene-C. Silk
samples were loosely affixed by sewing to temperature-
resistant coarse polyester mesh which allowed both free air
flow and also avoided contact between the metal and the
fabric surfaces during the thermal aging tests.

Light Exposures. Silk fabric samples, 0.25 m x 0.17 m,
were mounted in Atlas Electric Devices aluminum sample
holders according to AATCC Test Method 16E-1982 (7). An
Atlas Ci-35 Weather-Ometer xenon-arc was used on continuous
light cycle. Exposures were conducted at an irradiance of
0.35 W/m² measured at 340 nm and the irradiance was
monitored and controlled automatically. Borosilicate inner
and outer filters were used to simulate the solar spectrum.
The relative humidity was maintained at 65% and the black
panel temperature was 50°C. The actual fabric temperature
during the irradiation was measured, using small
thermocouples threaded into the fabric, and was found to be
35°C. Control samples for these tests were kept in the
dark at 35°C and 65% RH for the same time period as the
illuminated samples.
Samples of coated and uncoated silk fabric were removed
at various intervals up to an exposure of 605 kJ at 340 nm.
Maximum exposure of coated cloth was 242 kJ (340 nm). In
addition, for a fixed total energy of 86 kJ (340 nm), the
irradiance was maintained at 0.42 W/m², 0.35 W/m², and 0.28
W/m² to evaluate reciprocity effects. Samples of both types
of historic silk were exposed to 100 kJ (340 nm) and 360 kJ
(340 nm) to induce additional deterioration. These highly
exposed samples were coated with both 1 and 5 μm of
Parylene-C.
The spectral dependence of the light sensitivity (as
indicated by yellowing) of free films of Parylene-C was
determined. A Heraeus Sun-Test chamber, equipped with a
xenon arc lamp filtered to yield a simulated solar
spectrum, was used for the irradiation. An additional
infrared filter minimized sample heating. The irradiance
at the sample location was originally 0.83 W/m² at 340 nm,
but the output decreased approximately 20% after 1500 hours
use. Long band-pass optical filters with nominal cut-offs
of 305 nm, 345 nm, 385 nm and 400 nm were inserted between
the xenon lamp and the Parylene-C film samples to determine
the wavelength threshold for yellowing. The sample
temperature was maintained at 30± 2 °C with a water-cooled

support plate. Films were removed for color measurements
at intervals during a 500 hour exposure. The relative
spectral output and irradiance of the light source at the
sample site was calculated from the measured spectral
distribution and irradiance of the 1500 W xenon arc lamp at
the sample site (determined by DSET, Phoenix, Arizona) and
the transmission of the filters as measured with a
calibrated NBS tungsten lamp and an EG&G Gamma Scientific
Model 550 spectroradiometer.

Property Measurements. All silk samples were conditioned
prior to testing at 65% RH and 21°C according to ASTM Test
Method D 1776-79, "Standard Practice for Conditioning
Textiles for Testing." (8) Tensile properties were
determined on an Instron Model 4201 Universal Testing
Instrument. Tensile test data were recorded and stored for
reanalysis using Instron software, "General Tensile Test,
Revision D."
     For the modern silk fabric, 0.0254 m wide test samples
were cut using a Thwing Albert precision cutter. The gauge
length was 0.076 m and the constant rate of extension was
0.060 m/min. Five replicate samples were tested in accord
with ASTM Test Method D1682-64 (75), "Breaking Load and
Elongation of Textile Fabrics" (9). Because the historic
silk was more brittle than the modern silk, the extension
rate was reduced to 0.020 m/min. Owing to the limited
amount of cream-colored historic silk fabric available, the
sample width was reduced to 0.0125 m.
     Fabric color was determined with a Diano Matchscan 2
spectrophotometer. The reflectance spectra from 380 to 700
nm were measured over three layers of unexposed silk fabric
using a small area of view (SAV). CIELAB color coordinates
(L*,a*,b*) were calculated with the Matchscan software.
Color measurements on Parylene-C film were determined with
a Minolta Chromameter 221, a colorimeter with output
limited to CIE chromaticity or tristimulus values and
CIELAB L*, a* and b* color coordinates. Measurements on the
films after various exposure times were recorded with the
sample mounted over the white calibration plate.

Calculations. The stress-strain curves for the silk fabric
were plotted automatically from the data obtained with the
Instron. The initial modulus was determined from a
suitable straight line portion of the stress-strain curve.
The strain-to-break was then calculated with an effective
gauge length determined from extrapolation of the initial
modulus. The energy-to-break was calculated from the
integrated area under the corrected stress-strain curve to
the break point.
     The CIELAB values ΔE and hue angle were calculated from
L*, a* and b* values by using the following relationships:

$$\Delta E = (\Delta L*^2 + \Delta a*^2 + \Delta b*^2)^{0.5} \tag{1}$$

$$h_{ab} = Tan^{-1}(-b*/a*) \tag{2}$$

Results and Discussion

Initial Characterization. Modern silk was subjected to
tensile testing to evaluate the test procedure to be used,
the number of samples required and the expected precision.
Five fabric samples tested in the warp direction were found
to give a coefficient of variation of less than 5% for
breaking-load and less than 10% for strain-to-break.
    One of the more interesting initial differences between
the tensile properties of coated and uncoated fabric, shown
in Table 1, is the decrease in the initial modulus. The
Parylene-C coating used in the majority of the tests, 0.75
μm, was the smallest thickness at which a qualitative
change in the handle of this type of fabric was noticed.
In general, however, the handle is not of great
significance when considering the display of fragile silk
objects in a museum.

    Table 1. Tensile Properties of Coated and Uncoated
              Modern Silk Fabric

Property Change	Uncoated Mean	S.D.	Coated Mean	S.D.	%
Breaking-load	246	± 9 N	274	± 15	+11
% Strain-to-break	13.1	± 0.4	15.1	± 0.4	+15
Energy-to-break	0.147	± 0.007J	0.184	± 0.012J	+25
Initial Modulus (relative/area)	2450	± 80	2205	± 88	-10

    To determine the effect of coating thickness on tensile
properties, coating thicknesses of 0.25, 0.5, 0.75, 0.85,
1.0, 1.25 and 1.5 μm of Parylene-C were applied to modern
silk fabric. In Table 2, the breaking-load and strain-to-
break are shown as a function of coating thickness, and a
linear dependence is evident. A similar effect was
observed when paper was coated with Parylene-C (5).

    Table 2. Tensile Properties of Parylene-C Coated silk

Coating Thickness (μm)	Breaking-Load mean(N)	Strain-to-Break mean(%)
0.0	246	13.1
0.25	257	13.8
0.5	271	15.4
0.75	269	15.1
0.85	280	15.8
1.25	294	16.5
1.5	298	16.2

Thermal Exposures. The thermally induced changes in
tensile properties of coated and uncoated silk fabric,
expressed as percent retained breaking-load, strain-to-
break and energy-to-break in Figures 2,3, and 4,
respectively, are shown with the lines representing the
calculated exponential decline.
     To determine the kinetic order of the tensile property
degradation process,equations for both linear (Equation 3)
and exponential (Equation 4) behavior were used to fit the
data using the BMDP statistical computer program.

Zero order: $P_t/P_o = -kt$ (3)

First order: $P_t/P_o = e^{-kt}$ (4)

     Comparison of the residual sums of squares (RSS) for
the two models, Table 3, indicated that the exponential fit
is the more appropriate correlation and that the
degradation process for this fabric is first order.

Table 3.   Thermal Degradation Rate Constants
           for Modern Silk Fabric

Property				$k \times 10^4$, days^{-1}				
					Temperature,°C			
Coating	Fit	RSS	70	80	90	110	150
Breaking-load								
Uncoated	Linear	0.14	13	22	33	120	1080
	Exponential	0.09	14	25	43	160	1790
Coated	Linear	0.03	15	24	35	140	–
	Exponential	0.02	16	27	43	190	–
Strain-to-break								
Uncoated	Linear	0.21	19	34	39	140	1310
	Exponential	0.09	21	42	50	200	2630
Coated	Linear	0.05	28	34	42	460	–
	Exponential	0.02	35	29	54	240	–
Energy-to-break								
Uncoated	Linear	0.61	28	49	62	210	1600
	Exponential	0.21	34	69	100	380	4700
Coated	Linear	0.16	36	52	63	220	–
	Exponential	0.05	47	69	96	450	–

     Within the standard deviations of the rate constants,
k, (approximately 10%), it can be seen that: 1) the rate
constants for the changes in tensile properties are
different for each property at a constant temperature; 2)
degradation rates increase with temperature at different
rates for each property; and, 3) Parylene-C has no effect
on silk tensile property degradation rates.
     The last conclusion must be qualified because Parylene-
C exhibits an exponential rise in the permeability to gases
with a linear increase in temperature (10).   The
permeability to water vapor, for example, is increased by a

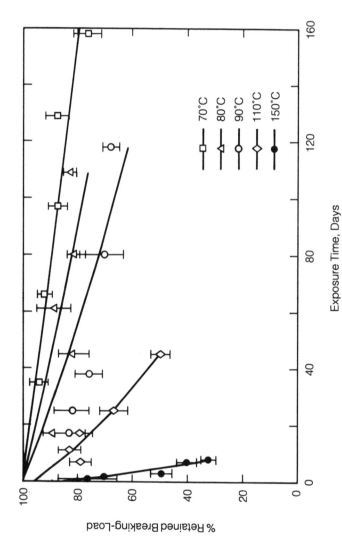

Figure 2. Effect of time at constant temperature on breaking-load of uncoated silk broadcloth.

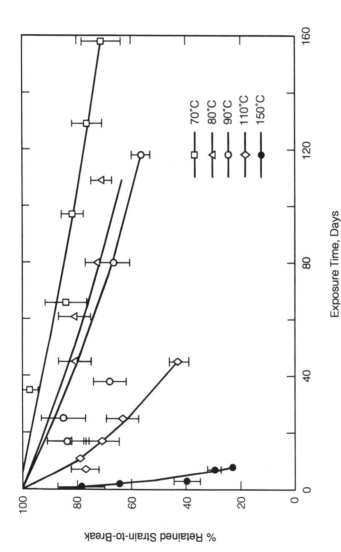

Figure 3. Effect of time at constant temperature on strain-to-break of uncoated silk broadcloth.

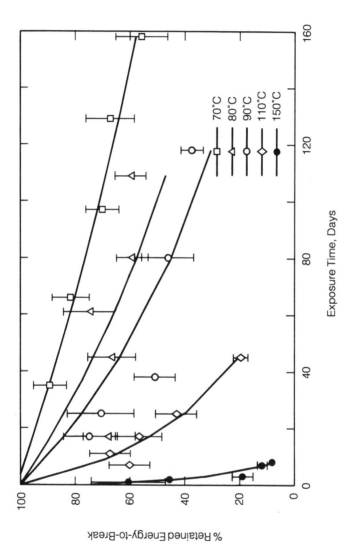

Figure 4. Effect of time at constant temperature on energy-to-break of silk broadcloth.

factor of 100 on raising the temperature from 20°C to 70°C.
For this reason, a coating of Parylene-C may have some
influence on the degradation rate at room temperature by
being an effective physical barrier to water vapor, a
possible effect that can not be taken into account in the
extrapolation of kinetic data through the use of an
Arrhenius relationship. Note, however, that the strain-to-
break and energy-to-break rate constants at 70°C for coated
and uncoated silk are different and outside the 10% error.

Estimation of tensile property degradation rates at
room temperature (20°C) from the high temperature kinetic
data was accomplished by application of the Arrhenius
equation:

$$k = A(e^{-Ea/RT}) \tag{5}$$

Coated silk was not tested at 150 °C because Parylene-C
undergoes a markedly increased degradation when heated
above 135°C. Because initial observations on the
discoloration of silk fabric indicated a change in the
nature of the degradation reaction at temperatures above
110°C, calculations were performed both including and
excluding the 150 C data.

Table 4.   Tensile Property Degradation Rates at 20°C
           Derived from the Arrhenius Relationship

Exposure Temperature Range Property	lnA	Ea,kJ	$k_{20}x10^2yr^{-1}$	$(t_{1/2})_{20}yr$
Uncoated, 70°C - 150°C				
Breaking-load	19.2	74.0	0.51	135
Strain-to-break	19.2	72.9	0.81	85
Energy-to-break	20.2	74.3	1.21	57
Uncoated, 70°C - 110°C				
Breaking-load	16.6	66.4	0.85	81
Strain-to-break	14.7	59.5	2.16	32
Energy-to-break	17.0	65.5	2.59	27
Coated, 70°C - 110°C				
Breaking-load	17.2	67.7	0.93	75
Strain-to-break	13.8	56.7	2.80	25
Energy-to-break	16.2	61.9	3.72	19

As can be seen in Table 4, the "activation energy"
derived from the series of exposures that includes silk
fabric heated at 150°C predicts a half-life for the strain-
to-break of 85 years at 20°C. The half-life at 20°C
calculated from the series of exposures that covers the
70°C to 110°C range is markedly different, only 32 years.
Such a large effect produced by inclusion of only one data
point indicates that results obtained at this temperature
may be suspect. By definition, the activation energy must
be independent of temperature over the range of
extrapolation.

It is important to remember that in the calculation of property degradation kinetics, the constants Ea and A do not imply any information regarding the chemical processes occurring. Multiple chemical reactions may give rise to an effective "activation energy" that is useful for comparison purposes only.

Tensile data obtained on silk fabric as a function of relative humidity at a fixed temperature level and time interval are shown in Figure 5. Between 0% and 50% RH, very little measurable change in properties was observed. Above 50% RH the degradation rates increased rapidly with increasing RH. Parylene-C also had no effect on the degradation rates of silk exposed at these RH levels.

The implications of these results are: 1) that the thermal aging of stored silk is minimized if the RH is lower than 50%; and 2) that reduction of the RH much below 50% will not reduce degradation rates appreciably. These results are applicable only in the absence of light.

CIELAB h_{ab} and ΔE were calculated from CIELAB color coordinates data for thermally aged coated and uncoated silk fabric. Figure 6 shows a plot of ΔE for the uncoated silk fabric as a function of exposure time at different temperatures. A linear regression analysis was applied to the data using the SPSS statistical computer program. The coefficient of determination (r^2) was greater than 0.9 in all cases. No significant differences were observed in the color change rate for coated and uncoated fabrics. There is no indication that the Parylene-C coating yellowed or affected the discoloration rate of silk in the temperature range studied.

A plot of CIELAB a* versus b* data for uncoated silk fabric exposed at 110°C and at 150°C is shown in Figure 7. The data obtained at 110°C are representative of all data obtained from 70-110°C and give a hue angle of $91\pm1°$. However, the final hue angle for the 150°C data was 78°, indicating a hue change toward the red, an effect that was not observed at lower temperatures. Furthermore, for a retained tensile strength of 70% and 48%, achieved by heating the fabric for different times at 150°C, the ΔE values were 24 units and 30 units, respectively. For the similar retained tensile strengths of 68 % and 52 %, achieved by heating at 110°C, the ΔE values were 8 and 11 units, respectively. Thus, the color developed in silk after exposure at 150°C was different not only in hue, but also in the extent of color change when compared with the color developed after heating silk samples in the 70°C to 110 C temperature range.

No attempt was made to apply the Arrhenius equation to the color change data because of the relatively large (±1 ΔE unit) variation of the color uniformity ("evenness") of the original uncoated silk fabric. At the lower temperatures, 70°C and 80°C, the measured color changes were close to the uncertainty in the color of the unexposed silk fabric itself and therefore too few temperature data points were available for meaningful calculations.

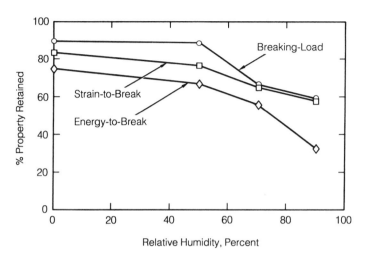

Figure 5. Effect of relative humidity on tensile properties of uncoated silk broadcloth.

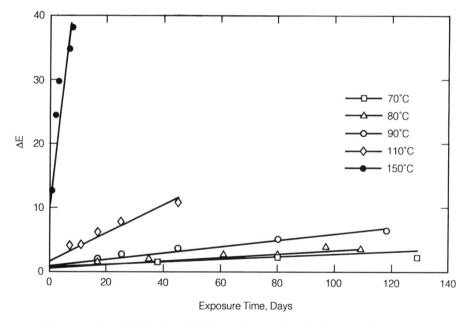

Figure 6. Effect of time at constant temperature on color change (ΔE) of silk broadcloth.

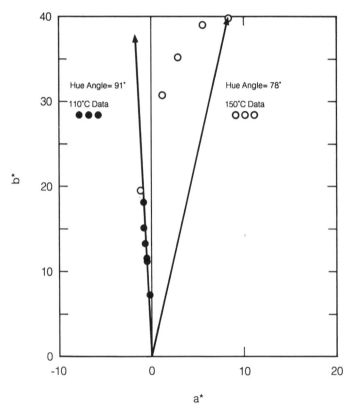

Figure 7. Hue angle of silk broadcloth at different temperatures.

The effect of humidity variation on the color change of uncoated silk fabric at constant temperature (90°C) for a fixed exposure time (17 days) is shown in Figure 8. The rate of change in color increases with relative humidity above 50 % for both coated and uncoated fabric. This is the same humidity level above which the change in the tensile property degradation rates also increases.

Light Exposures. The tensile property changes of uncoated silk fabric following exposure to light aging are shown in Figure 9-11. First-order and zero-order degradation rates are indistinguishable for uncoated fabric after an exposure of 242 kJ (340 nm). A closer fit of the data to an exponential rather than a linear model is only evident after an exposure of 605 kJ. The breaking-load was found to be the least sensitive of the tensile properties and therefore should not be used as the only diagnostic tool in studies of the photolytic degradation of silk.

Initially, the tensile properties of Parylene-C coated silk cloth were higher than that of the uncoated cloth (with the exception of the initial modulus). However, after a 242 kJ exposure to the xenon-arc source, the magnitude of the tensile properties of both materials were the same.

No significant difference in the degradation rate of tensile properties for either uncoated or coated silk was found on variation of the irradiance level of the xenon lamp from 0.28 W/m² to 0.42 W/m² (340 nm) for a constant total energy input of 86 kJ (340 nm). Apparently, reciprocity holds in this range of irradiances.

Uncoated silk fabric yellowed on exposure to light. The color change, shown in Figure 12 as CIELAB ΔE, reached a maximum of 5 units after a 50 kJ (340 nm) exposure and remained constant on further exposure. Parylene-C coated silk continued to yellow out to an exposure of 242 kJ (340 nm).

In contrast to the color changes that occurred in modern, uncoated silk after exposure to heat, the light-induced color change was not indicative of the tensile properties of the fabric. Loss in tensile properties with irradiation continued while the color remained static. This behavior was not the same for Parylene-C coated, fabric which continued to yellow with irradiation and also continued to exhibit a decline in tensile properties to 242 kJ (340 nm).

Over the range of irradiance from 0.28 W/m² to 0.45 W/m² (340 nm) for a total exposure of 86 kJ, the change in color for either coated or uncoated fabric was the same and was independent of irradiance.

Effect of wavelength on the yellowing of Parylene-C films. The change in color of 12.5 µm-thick films of Parylene-C following irradiation under a set of long pass optical filters is shown in Figure 13. The normalized bar plot (Figure 14) was obtained by dividing ΔE by the power

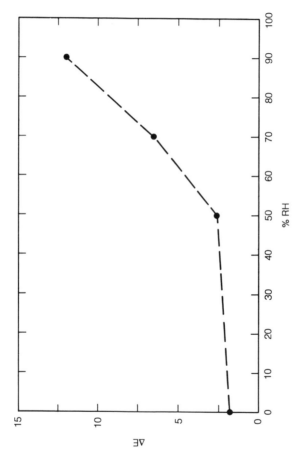

Figure 8. Effect of relative humidity on color change (ΔE) of uncoated silk broadcloth.

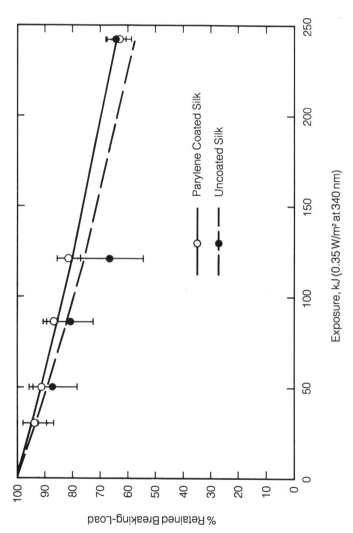

Figure 9. Effect of light on breaking-load of silk broadcloth.

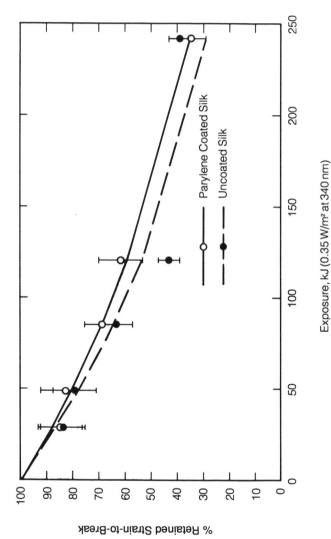

Figure 10. Effect of light on strain-to-break of silk broadcloth.

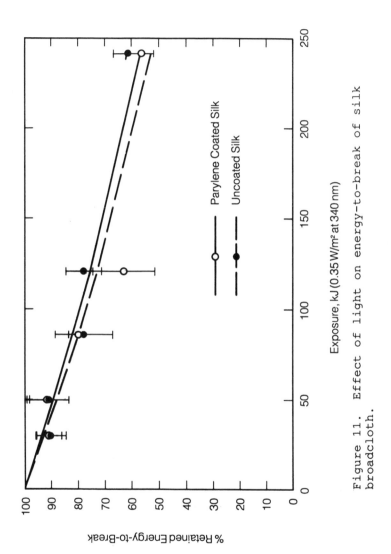

Figure 11.   Effect of light on energy-to-break of silk broadcloth.

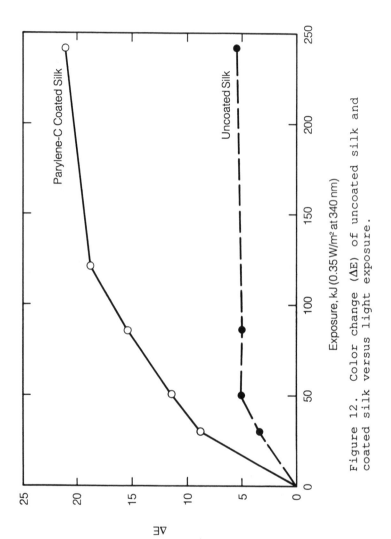

Figure 12. Color change (ΔE) of uncoated silk and coated silk versus light exposure.

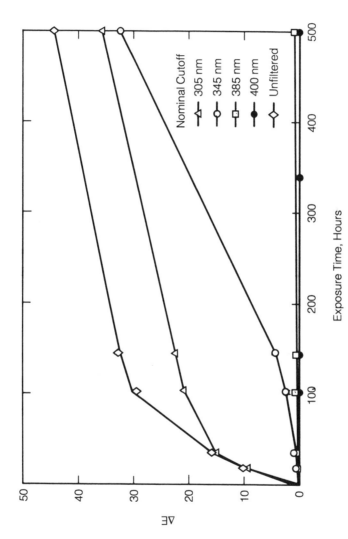

Figure 13. Color change (ΔE) of Parylene-C films versus light exposure.

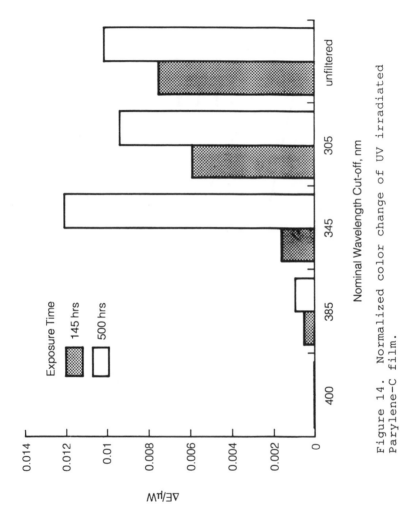

Figure 14. Normalized color change of UV irradiated Parylene-C film.

transmitted through the individual filters in the 300 nm to
400 nm range. It can be seen that significant yellowing
does not occur at wavelengths greater than 385 nm. The
long-pass filters exhibited transmission tails 15 nm below
the nominal cut-off, and therefore some radiation as low as
385 nm passed through the 400 nm filter.

Because the most common museum ultraviolet filters
presently in use have transmissions that are comparable to
the 385 nm nominal cut-off filter used in the tests,
yellowing would be expected. In order to suppress the
discoloration of Parylene-C-coated artifacts, a filter with
an effective cut-off at 400 nm is required. After a 500
hour exposure, the color change of Parylene-C exposed
through the 385 nm filter was 1 ΔE unit.

Effect of Parylene-C Coating on Historic Silk Fabric. The
results of tensile testing of coated and uncoated historic
silk fabric are given in Table 5. For modern, strong silk
(original, uncoated breaking-load >250 N, strain-to-break
>15%), a 0.75 μm thick layer of Parylene-C increased the
tensile properties, with the exception of the initial
modulus, as was shown in Table 1.

Table 5.   Tensile Properties of Uncoated and Coated
Historic Silk

Exposure Coating,	Breaking-load, N	% Strain-to-break	Energy-to-break, J	Initial Modulus Relative/area
Cream-colored silk				
Unaged				
0	50 ±5	11.4 ±0.9	0.192 ±0.04	5600 ±500
1	64 ±3	11.4 ±0.4	0.213 ±0.04	6600 ±460
5	83 ±4	11.4 ±1.0	0.219 ±0.04	17800 ±1400
Heat aged, 30 days				
0	3 ±1	2.3 ±0.5	0.001	2800 ±220
1	8 ±2	2.7 ±0.4	0.004 ±0.002	3400 ±1500
Light aged, 350 kJ (340 nm)				
0	7 ±1	4.4 ±0.3	0.008 ±0.002	2700 ±200
1	7 ±0.5	4.1 ±1.0	0.009 ±0.002	3600 ±360
Blue-Silk				
Unaged				
0	32 ±4	5.4 ±0.5	0.041 ±0.006	11000 ±1400
1	42 ±9	3.9 ±0.4	0.049 ±0.012	17700 ±2300
5	65 ±12	2.9 ±0.2	0.065 ±0.012	38800 ±1300
Heat aged, 10 days				
0	9 ±3	1.8 ±0.4	0.006 ±0.002	10500 ±2600
1	16 ±3	2.7 ±0.6	0.009 ±0.002	16700 ±2000
5	35 ±13	1.9 ±0.28	0.012 ±0.006	38000 ±4000
Light aged, 100 kJ (340 nm)				
0	2.6±1.2	1.9 ±0.3	0.003 ±0.001	7800 ±1000
1	6 ±3	1.9 ±0.5	0.005 ±0.002	18700 ±1200
5	18 ±9	1.77±0.8	0.006 ±0.003	30600 ±3000

For the cream-colored fabric, coating with Parylene-C
resulted in both increased initial modulus and an increased
breaking-load. However, no increase in the strain-to-break
was observed. This was found to be the case for both 1 and
5 μm coating thicknesses.
     For the blue fabric, the results showed an increased
breaking-load upon coating, but a decreased strain-to-
break, which resulted in an actual reduction in the ability
of the fabric to elongate and absorb shock, and therefore,
little increase in the energy-to-break. However, the
initial modulus also was increased.
     Historic fabric samples that were further deteriorated
by heat or light exposures prior to coating with Parylene-C
showed the same trends: increased breaking-load and initial
modulus with no added strain-to-break. The absolute
increase in breaking-load was small when compared with the
increase that resulted when modern silk fabric was coated,
but the percentage increase was higher due to the initially
low absolute values of the tensile properties.
     A large increase in the absolute tensile property
values of the Parylene-C coated silk fabric was only
apparent when the properties of the fabric were initially
large. Brittle silk fabric does not become more elastic as
a result of being coated with a thin layer of Parylene-C.
The effect on the energy-to-break is thus not great for a
fabric that, before coating, has little ability to
elongate. However, for a fabric on display or in museum
storage, this may not be an essential requirement.

Conclusions

     It has been shown that the application of a thin
(0.75μm), vapor-deposited coating of Parylene-C on modern
silk broadcloth increased the tensile properties, with the
exception of the initial modulus, and did not alter the
color or appearance of the fabric. Coating brittle,
degraded historic silk resulted in an increased breaking-
load and initial modulus but did not appreciably improve
the pliability as indicated by the strain-to-break. The
coating should then be considered, in the conservation of
degraded silk, primarily for a situation where increased
tensile strength is the major consideration, such as static
display with only lateral forces in play.
     The Arrhenius relationship of the property degradation
rates of uncoated silk over a range of temperatures was
determined. A temperature of 150°C was found to be
inappropriate for the accelerated thermal aging of silk
because of the large effect on the extrapolated reaction
rate at 20°C for tensile properties and also because of the
amount and hue of color developed. In addition, changes in
the relative humidity affected the degradation rate of
tensile properties and yellowing only above a level of 50%.
     The coating of Parylene-C did not have an effect on the
thermal degradation rates of the silk broadcloth tensile
properties or yellowing. The coating deteriorated when

exposed to a light source simulating the outdoor solar
spectrum. Elimination of ultraviolet light to the extent
of a 400 nm nominal cut-off was necessary to suppress the
discoloration of free-films of Parylene-C. Display of
coated artifacts should be limited to similar conditions of
illumination, either by choice of a filter or choice of
illuminant. However, the effect on the tensile properties
of coated silk fabric exposed to light filtered from
ultraviolet radiation was not measured.

Acknowledgments

The authors wish to thank Dr. Terry Reedy for the
statistical analysis; Max Salzman, who arranged for the
sample of historic cream-colored silk from the Smithsonian
Institution; and, John Twilley, who supplied the sample of
historic blue silk from the Los Angeles County Museum of
Art.

Legend of Symbols

A	Arrhenius constant
a^*	CIELAB redness-greenness coordinate
b^*	CIELAB yellowness-blueness coordinate
C	Temperature, Celsius
ΔE	CIELAB color change
Ea	Activation energy
h_{ab}	CIELAB hue angle
k	rate constant
k_{20}	rate constant at 20°C
L^*	CIELAB lightness-darkness coordinate
P_t	Property value at time t
P_o	Initial property value, t=0
R	Gas constant
t	Time
$(t_{1/2})_{20}$	Half-life at 20°C
T	Temperature, Kelvin

Literature cited

1   Hirabayashi, H., Arch. and Nat. Sci., **1984**, 17, 361.
2   Scott, W., Am. Dyestuff Rep., **1931**, 20, , 583.
3   Verdu, J.; Kleitz, M.,IIC Pre., Paris Con., **1984**, 64.
4   Masschlein-Kleiner, L.; Bergiers, F., IIC Pre., Paris
    Con., **1984**.
5   Humphrey, B., JAIC, **1986**, 25, 15.
6   Humphrey, B., Stud. Cons., **1984**, 117.
7   "AATCC Technical Manual"; American Association of
    Textile Chemists and Colorists: Research Triangle Park,
    North Carolina, **1982**, 17, 160-162.
8   Annu. Book ASTM Stand. **1987**, 07-02, 417.
9   Annu. Book ASTM Stand. **1987**, 07-01, 343.
10 Nowlin, T., J. Poly. Sci.: Poly. Chem., **1980**, 18,2013.

RECEIVED February 28, 1989

Chapter 9

Historic Silk Flags from Harrisburg

M. Ballard[1], R. J. Koestler, C. Blair, C. Santamaria[1], and N. Indictor[2]

Department of Objects Conservation, Metropolitan Museum of Art, New York, NY 10028

A group of thirty-four silk samples taken from fourteen brittle Civil War flags was provided by the Capitol Preservation Committee, Harrisburg, Pennsylvania. Elemental analyses of these samples were obtained by energy dispersive x-ray spectrometry (EDS). Total sulfur, ash, and pH of the samples is also reported, and colorants were determined. The presence of mordants, weighting materials, and colorants is discussed with reference to the embrittlement of the silk. The connection between fiber deterioration and color is discussed as well as the effects of contemporary manufacturing treatments on the present fiber condition. None of the samples examined was weighted with inorganic salts.

The phenomenon of silk fabric degradation in dark storage from the action of mineral salts used in the processing of the silk, especially the 19th and 20th century use of tin salts as weighting agents, has been a concern for textile curators, conservators and historians for many years(1-4). Standard test methods used for the assay of weighting materials generally requires larger samples than can be sacrificed from historic textiles. Ordinary scanning electron microscopy (SEM) seems to fail as a method for distinguishing weighted from unweighted samples (5), but recently the use of transmission electron microscopy (TEM) on uncoated single fibers has shown what appears to be clear differences (6). As shown earlier, satisfactory results may be obtained by energy dispersive x-ray spectrometry (EDS) supplemented by ash content determinations. An initial study on controlled modern silk samples involved the detection of tin and iron weighting in various recipes chosen form the literature (5).

With this experience, small samples taken from flags in the National Museum of American History's Division of Armed Forces History were analyzed (1). The flags dated from the Civil war through World War I. The results of analysis showed tin-weighting in only one of the degraded silk flag samples - a French gift to General Pershing

[1]Current address: Conservation Analytical Lab, Smithsonian Institution, Washington, DC 20560
[2]Current address: Chemistry Department, Brooklyn College, and the Graduate Center, City University of New York, Brooklyn, NY 11210

during World War I. A consistent feature of that sample set was a
sulfur content higher than could be accounted for from the sulfur
containing amino acids in silk, and probably higher than any sulfur
that may have been part of the colorants. Only cysteine and
methionine in Bombyx mori contain sulfur (total S= $\underline{ca}$ 0.05-0.07%)
($\underline{7,8}$). Early synthetic dyes with sulfonic acid groups were indicated
by spot tests on some of the colorants; but sulfur was prominent in
dyed and undyed samples alike.

English, French, and German 19th century dye texts indicate that
many silks were not so much tin weighted as they were 'brightened' by
a final 'sour' or rinse in sulfuric acid. White silk tended to be
bleached of any original yellowness with SO_3 ($\underline{9-13}$). Accretions of
dirt, soil, or atmospheric pollutants on the surface may also account
for some of the sulfur. The War Department donated these flags to the
Smithsonian Institution in 1919. It is known that some of the flags
held by the War Department were in poor condition as early as the
1880's; the entire collection was shipped to the Smithsonian
Institution by boxcar ($\underline{14}$).

In the present study, a well documented collection of samples
was examined (Table I, Figures 1 and 2). The Capitol Preservation
Committee, Harrisburg, Pennsylvania provided samples of some Civil
War flags, manufactured mostly by Horstmann Brothers & Company or
Evans & Hassall. These flags had been sealed in glass cases beneath
individual silk chiffon sleeves on their hoists form 1901 until a
conservation program was organized in the early 1980's. In addition
to conservation documentation, a full scale historical review of the
flags has been carried out and published ($\underline{14}$).

Experimental.

EDS Analyses. Sample preparation and data treatment has been
described ($\underline{1,5}$). Photomicrographs and printouts of EDS scans were
retained for files.

Ash and Total Sulfur Analyses. Samples were removed from the
specimens and submitted for ash and total sulfur analysis to
Schwartzkopf Microanalytical Laboratory, Woodside, N.Y. 11377. For
ashing, samples were burned under oxygen at 900-1000°C for $\underline{ca}$ one
half hour. For sulfur, samples were treated with potassium metal to
convert sulfur to sulfides, followed by conversion to methylene blue
for spectrophotometric analysis (at 670 nm). Except for very small
samples (less than 0.2mg), values for ash and sulfur have an
uncertainty of $\underline{ca}$ ±10% of the reported values.

pH and Color Analyses. Samples 1 x 1 cm^2 were placed in test tubes
with 3-4 ml deionized water. pH of the water was measured after five
minutes at ambient temperature using a Corning Model #12 pH meter and
a Markson combination electrode. (The pH of modern silk samples
measured in the same way was $\underline{ca}$ 5.9±0.1). The same samples used for
pH measurements were analyzed for colorants according to procedures
of H. Schweppe ($\underline{15,16}$)

Table I. Description of Flags Sampled

No.	Registra-tion No.	Regiment	Historical Name	Manufacturer
012	1985.012	23rd PA	State Color	Evans & Hassall
019	1985.019	28th PA	State Color	Evans & Hassall?
022	1985.022	29th PA	State Color	Horstmann Bros. & Co.
026	1985.026	31st PA (2nd res)	State Color	Horstmann Bros. &Co.
096	1985.096	63rd PA	State Color?	Evans & Hassall?
137	1985.137	82nd PA	State Color	Unknown
175	1985.175	Unknown	State Color	Evans & Hassall?
066	1985.066	51st PA	State Color	Horstmann Bros. & Co.
167	1985.167	51st PA	Guidon, National	Unknown
168	1985.168	51st PA	Guidon, National	Unknown
005	1985.005	59th PA (2nd Cav)	State Standard	Horstmann Bros. & Co.
050	1985.050	44th PA (1st Cav)	State Standard	Unknown
002	1985.002	78th PA	National Regimental	Unknown
103	1985.103	67th PA	National Regimental	Unknown

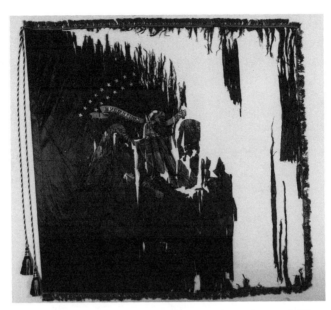

Figure 1. 1985.002; 78th PA National Regimental Flag.

Figure 2. 1985.019; 28th PA Regiment State Color.

Results and Discussion

Table II lists the EDS results of the thirty-four specimens
qualitatively. Only elements of atomic number greater than 11 are
sensed in the EDS analyses. The results were obtained an normalized
percentages using a software program (ASAP, a Kevex proprietary
procedure) usually employed in analyses from this laboratory. The
signs used in Table II should be interpreted as follows:

++ = The element is present in significant amounts (more than
30%, of elements of atomic number greater than 11). The element
is found in all scans.
+ = The element is present in all scans (more than ca 5%) of
elements of atomic number 11).
(+) = The element is present in most, but not necessarily all
scans (ca 5% of elements of atomic number greater than 11).
- = The element is not detected.

The wide variety and variability of elements detected for
identically colored silks suggests, as previously observed, that the
methods of coloring these historical flags was not standardized.
Every sample showed the presence of iron. Of the thirteen blue
specimens eleven contained tin; two did not. Of the ten red samples,
eight contained tin; two did not. Only two of the ten white samples
contained tin. Sulfur, calcium, silicon, aluminum were found in all
samples; chlorine, potassium, and sodium were frequently found;
magnesium was detected occasionally in trace quantities.
    Table III lists the results of pH measurements, ash, and total
sulfur analyses grouped according to sample color.
    All pH measurements showed lower values, (usually by more than
2.0 pH units), than were found for modern silk samples suggesting
rather clearly the occurrence of chemical changes in the historic
silks. Blue samples had the highest pH, average ca 4.8±0.5; red
average ca 4.2 ±0.3; white (three) ca 4.1±0.3.
    Except for a single example (012) the sulfur analyses were
substantially higher than values reported for untreated silk, i.e the
sulfur accountable to the presence of cysteine or methionine in the
fibroin chain of silk. Values obtained for total sulfur ranged from
ca 2 - 9 times the value expected from untreated silk. The single low
value found for a blue sample was very close to that expected for
untreated silk; the highest value was for a white sample. Most of the
samples (twenty-five of thirty-four)had a sulfur % of 0.35±0.12, ca 5
times the amount present in untreated silk.
    The ash content of the samples indicated clearly that none of
the samples was weighted with inorganic substance. The clearest
property distinction among the colors was observed in the ash %
results: The blue samples gave the highest values, average ca
4.9±0.4%; red, average ca 3.2±1.0%; white, average ca 1.6±0.3%. These
average values were calculated neglecting data from small samples
(Table III, date with *). Although there were a few overlapping
results, for the most part there seems to be a high correlation
between color and ash content.
    The colorants for all the flags was found to be consistently
cochineal for the reds and Prussian Blue for the blues. Although the

Table II. Qualitative EDS Results

Sample No.*	Na	Mg	Al	Si	S	Cl	K	Ca	Fe	Sn
012R	+	-	(+)	+	+	(+)	+	+	+	-
Y	(+)	-	(+)	+	+	(+)	+	+	(+)	-
B	(+)	(+)	(+)	(+)	+	(+)	-	+	++	++
019R	(+)	(+)	(+)	+	+	-	(+)	+	+	+
Y	(+)	-	(+)	+	++	(+)	(+)	+	(+)	+
B	(+)	(+)	(+)	+	+	-	-	+	++	++
022R	(+)	-	(+)	+	+	-	(+)	+	+	+
Y	(+)	-	(+)	+	+	(+)	+	+	+	-
B	(+)	-	(+)	(+)	+	(+)	-	+	++	++
026R	(+)	-	+	+	+	-	+	+	+	(+)
Y	-	-	+	+	+	(+)	+	+	+	-
B	(+)	-	(+)	+	+	-	+	+	++	+
096R	(+)	-	+	+	+	+	+	+	+	(+)
Y	(+)	-	+	+	+	+	+	+	+	-
B	(+)	-	(+)	+	+	(+)	+	+	++	++
137R	(+)	-	+	+	+	(+)	+	+	+	-
Y	(+)	-	(+)	+	+	(+)	+	+	+	-
B	-	-	(+)	+	+	(+)	-	+	++	++
175R	+	-	+	++	+	+	+	+	+	(+)
Y	-	-	+	+	++	(+)	+	+	+	-
B	(+)	-	(+)	+	+	(+)	-	(+)	++	++
066R	(+)	-	+	+	+	(+)	-	+	+	++
Y	(+)	-	+	+	+	(+)	+	+	+	-
B	(+)	(+)	(+)	(+)	+	(+)	-	(+)	++	++
167R	(+)	-	(+)	+	+	-	-	+	(+)	++
Y	-	-	(+)	+	++	(+)	-	+	+	+
B	-	-	(+)	++	++	(+)	(+)	+	+	-
168R	(+)	-	(+)	+	+	(+)	+	+	+	+
Y	-	(+)	(+)	+	++	-	+	+	+	-
B	-	(+)	(+)	+	++	(+)	(+)	++	(+)	-
005B	(+)	(+)	(+)	+	+	-	(+)	(+)	++	+
030B	-	-	+	+	+	-	-	+	+	+
002B	+	-	(+)	+	+	+	(+)	+	+	+
103B	+	-	+	+	+	(+)	(+)	+	+	+

*RYB refers to color of samples
++ = Present in significant amounts (>30% of elements of atomic no. >11).
+ = Present. (*ca.* 5–30% of elements of atomic no. >11)
- = Absent (not detected).
(+) = Possibly present (*ca.* 5% or less of elements of atomic no. >11).

Table III. pH, Sulfur, Ash of Historic Flags

Sample No.	pH[a]	Sulfur[b] %	Ash[c] %
BLUE SAMPLES			
012	4.61	0.071	5.95
019	4.97	0.27	6.49
022	5.18	0.19	7.52
026	4.60	0.36	4.75
096	4.86	0.30	5.62
137	4.64	0.33	<0.1*
175	4.17	0.32	5.49
066	5.30	0.15	5.44
167	4.15	0.30	<0.3*
168	4.08	0.34	<0.6*
005	4.81	0.28	5.15
050	5.28	0.18	7.72
002	4.85	0.35	5.11
103	5.35	0.38	7.21
RED SAMPLES			
012	4.08	0.15	3.62
019	3.87	0.37	<2.8*
022	4.28	0.16	1.76
026	3.88	0.33	2.73
096	4.25	0.24	4.30
137	4.19	0.17	2.73
175	4.31	0.47	4.79
066	4.77	0.31	n.d.
167	3.86	0.30	2.28
168	4.29	0.23	<2.9*
WHITE SAMPLES			
012	n.d.	0.46	1.84
019	n.d.	0.28	0.83
022	n.d.	0.23	1.03
026	n.d.	0.42	1.66
096	4.41	0.39	3.65
137	n.d.	0.31	1.75
175	3.79	0.42	1.26
066	n.d.	0.15	1.33
167	3.97	0.31	1.25
168	n.d.	0.66	1.94

[a] Modified cold extraction.
[b] Samples burned under Oxygen at 900–1000° C, ca. 0.5 hr.
[c] Total sulfur: S converted to methylene blue, determined spectrophotometrically.
* = Sample < 0.2 mg
n.d. = no data

dye substances were the same for each color, the general
manufacturing processes, (washing, bleaching, mordanting, post dyeing
rinses) were probably not rigorously standardized. The analyses
described here were unable to distinguish the manufacturer.
Variations in ash, sulfur, and tin do not indicate
manufacturer-specific silk processing. In fact the variations
indicate a fairly broad range of possible manufacturing conditions to
achieve the products surveyed.

For conservators and curators the connection between the
condition of the object and the manufacturing technology is
important: first in that on some level, the history of its
manufacture defines an object; and secondly because the deterioration
of the object may be inextricably linked to its manufacturing
origins. Although is is not possible to correlate the present data
either with a manufacturer or processing, a number on interesting
generalizations may be made. For these specific flags it is quite
possible to reconstruct from the accumulated records, actual battle
use, subsequent souveniring, and long term display and storage.
Anecdotal evidence among textile conservators suggests that the
relationship between silk color and condition (white worse than red
worse than blue) should be an interesting subject for further
investigation.

It is clear that these flags, like the majority of the
deteriorated flags studied earlier form the National Museum of
American History contain no inorganic weighting agents. This study
has not addressed the possibility of organic weighting materials.
Sugar, gelatin, and tannins were also described as weighting agents
in nineteenth century texts (13).

Additional work on the preferential loss of amino acid residues
with ageing would shed light on fundamental changes that occur in
silk deterioration. Further work may suggest productive avenues for
conservation treatment. Many of the flags have maintained their
integrity as object 'in poor condition' since the 1880's. To date
there is no evidence that these objects, untreated in flat storage,
are actively disintegrating at an accelerating rate as a result of
tin weighting.

Summary

1. The deteriorated condition of historic silk flags appears not
generally to be the result of weighting with inorganic substances.
None of the samples studied showed evidence of weighting with
inorganic material.
2. The red colorant was cochineal in all examples of this group; the
blue colorant was Prussian Blue in all examples of this group.
3. EDS analyses showed the presence of sulfur, iron, calcium,
silicon, aluminum, in each sample analyzed; tin, chlorine, potassium,
sodium, were usually found.
    Blue samples: Twelve of fourteen contained tin
    Red samples: Eight of ten contained tin
    White samples: Two of ten contained tin
These variations suggest that no general procedure for processing the
silk was practiced, except for the selection of colorants.

4. Total sulfur analyses showed greater levels (2-9x) than could be
accounted for by the sulfur in the proteinaceous backbone of the
silk.
5. The pH of the blue samples was highest; the pH of the white
samples was lowest.
6. The ash content of the blue samples was highest; the ash content
of the white samples was lowest.

Literature Cited

1. Ballard, M.; Koestler, R.J.; Blair, C.; Indictor, N.; "A Study of
   Historic Silk Banners by Scanning Electron Microscope-Energy
   Dispersive X-Ray Spectrometry," Symposium on Preservation of
   Papers and Textiles of Historic and Artistic Value, 193rd American
   Chemical Society Meeting, Denver, Colorado, April 5-10, 1987.
   Preservation of Papers and Textiles of Historic and Artistic
   Value, Advances in Chemistry, American Chemical Society;
   Washington, D.C., (in press).
2. Howitt, F.O., Bibliography of Silk.; Hutchinson's Scientific and
   Technical Publications: London, 1946; Ch. 8.
3. Ross, J.E.; Johnson, R.L.; Edgar, R., Textile Res.J. 1936, 6,
   207-216.
4. Ballard, M.; Koestler, R.J.; Indictor, N., Preprints of Papers,
   13th Annual Meeting of the American Institute for Conservation of
   Historical and Artistic Works, Washington D.C., May 1985, p. 155.
5. Ballard, M.; Koestler, R.J.; Indictor, N., Scanning Electron
   Microsc. 2, 1986, 499-506.
6. Blair, C., unpublished results.
7. Handbook of Common Polymers, The Chemical Rubber Co.: Cleveland
   OH, 1971.
8. Stevens, M.P., Polymer Chemistry--An Introduction,
   Addison-Wesley: Reading MA, 1975.
9. Ganswindt, A., Dyeing Silk, Mixed Silk Fabrics and Artificial
   Silks, trans. by Salter, C.: Scott, Greenwood and Son: London,
   1921.
10.Hummel, J.J.., Colouring Matters for Dyeing Textiles, rev, ed,;
   Hasluck, P.N., Ed.: Cassell and Co., Ltd.: London, 1906.
11.Hummel, J.J.., Mordants, Methods,and Machinery Used in Dyeing,
   rev. ed.; Hasluck, P.N., Ed.; Cassell and Co., Ltd.: London, 1906.
12.Hurst, G.H., Silk Dyeing, Printing, and Finishing, George Bell
   and Sons: London, 1892.
13.Knecht, E.; Rawson, C.; Loewenthal, R., A Manual of Dyeing, 2nd
   ed.; Charles Griffin and Co., Ltd,; London, 1910.
14.Savers, R.A., Advance the Colors! Pennsylvania Civil War Battle
   Flags, vol. 1; Capitol Preservation Committee: Harrisburg, PA,
   1987.
15.Schweppe, H., Practical Information for the Identification of
   Early Synthetic Dyes, Conservation Analytical Laboratory,
   Smithsonian Institution: Washington, D.C. 1987.
16.Schweppe, H., Practical Information for the Identification of
   Dyes on Historic Textiles Materials, Conservation Analytical
   Laboratory, Smithsonian Institution: Washington, D.C., 1988.

RECEIVED February 22, 1989

Chapter 10

Long-Term Stability of Cellulosic Textiles

Effect of Alkaline Deacidifying Agents on Naturally Aged Cellulosic Textiles

N. Kerr, T. Jennings, E. Méthé

Department of Clothing and Textiles, University of Alberta, Edmonton, Alberta T6G 2N1, Canada

Research has shown that deacidifying agents are effective in reducing the degradation of new cellulosic textiles as well as paper, however, conservators are hesitant to deacidfy historic textiles. This research addresses their concerns about possible changes in fabric properties if historic textiles are deacidified. The effect of magnesium bicarbonate, calcium hydroxide and methoxy magnesium methyl carbonate (Wei T'o solution #2 and #12 spray) on physical properties has been studied. Thirteen naturally aged cottons and linens, 7 commercially dyed cottons and 3 cottons artificially aged by bleaching and ionizing radition were deacidified. Color change was minimal on the naturally aged cottons and dyed cottons, but the artificially aged cottons darkened noticeably. Moisture regain increased by 0.2 to 0.4% and weight gain was 2 to 5%. Four treated fabrics were subjected to accelerated aging at 100°C, and 100% RH for 15 days; the pH of these fabrics stayed above 6.6 while untreated specimens became noticeably acidic (pH 3.8-5.1). Fabrics stiffened considerably particularly when immersed in Wei T'o #2, but the stiffness did not reduce the flex abrasion of old fabrics. SEM photomicrographs show that the surfaces of fibers are essentially unchanged, but some fibers have adhered together. New methods of application are required to reduce stiffening.

0097–6156/89/0410–0143$06.00/0
© 1989 American Chemical Society

Researchers have shown that deacidifying agents are
beneficial in improving the longevity of cellulosic
textiles as well as paper (1-5). The development of
acidity in cellulosics through aging and exposure to
atmospheric pollutants promotes the degradation of these
fibers. Although it has been shown that deacidifying
agents can neutralize acidity in textiles and provide an
alkaline reserve, conservators are reluctant to use such
agents. This reluctance is justifiable to a certain
extent because historic artifacts are irreplaceable and no
treatment should be adopted until its long term effects
are known. In studying deacidifying agents, researchers
have frequently used new rather than naturally aged
fabrics exhibiting signs of oxidation and hydrolysis.
While the alkalinity of a deacidifying treatment does not
harm new cotton or linen, there is fear that it will
damage a highly oxidized historic fabric and cause
yellowing (6-7). Changes in hand, drape and flexibility
may affect the appearance of a historic textile. It is
possible that abrasion resistance and moisture regain of a
treated fabric may be altered. The purpose of this
research is to address the above concerns conservators
have expressed regarding physical changes brought about by
deacidification treatments. These treatments must be
found acceptable, practical and workable from a
conservator's point of view, or else this potentially
valuable tool for extending the life of cellulosic
textiles may never be used.

Experimental

Fabrics. New fabrics, naturally aged fabrics and fabrics
aged by artificial means were used in this study. The
fabric used in the new state and after artificial aging
was desized, scoured, unbleached cotton (400U) purchased
from Testfabrics, Inc., Middlesex, NJ 08846; count of
35x31 yarns/cm (warp x weft) and weight of 117 g/m^2. This
fabric was subjected to accelerated aging by exposure to a
dose of 50 or 75 Mrads of ionizing radiation or by
chlorine bleaching. Irradiation in an electron
accelerator (High Voltage Engineering Corporation) was
carried out at the College of Textiles, North Carolina
State University. Although the mode of degradation
differs from natural aging, the end result is similar in
that the cellulose is both hydrolyzed and oxidized (8).
Fabric was aged by bleaching for 30 minutes at pH 7 with
hypochlorite solution prepared according to Burgess and
Hanlan (9). Bleaching under these conditions produces a
reducing oxycellulose with many carbonyl groups. Thirteen
naturally aged fabrics, varying in age from 25 to 90
years, were obtained from the University of Alberta
Historic Costume Collection, the Marian Centre and private
donors in Edmonton, Alberta (see Table I). A 100% cotton
fabric, commercially dyed in seven colors, was used to

examine the effect of deacidifying agents on the color of
dyed fabrics. New and naturally aged fabrics were washed
in a 0.1% anionic detergent solution and rinsed repeatedly
in distilled water before treatment with deacidifying
agents.

Table I.   Description of Undyed Naturally Aged
            Cellulosic Textiles[1]

Specimen Description	Color	Condition	pH
1 linen tea cloth	white	excellent	6.9
2 fine linen tea cloth	grey-white	many holes	6.7
3 linen runner	yellow-tan	many holes	6.7
4 cotton napkins	stained	good	6.5
5 cotton voile shirt, c. 1960	yellowed	good	5.4
6 cotton shirt starched	white	a few holes	6.9
7 light weight cotton	yellowed	weakened areas	5.8
8 cotton night gown, c. 1900	yellowed	good	5.5
9 cotton bed sheet, c. 1960	white	worn areas	-
10 cotton bed sheet, c. 1960	white	poor, very worn	-
11 cotton bed sheet, c. 1950	white	good	6.0
12 cotton bed sheet, c. 1960	yellowed	weakened areas	6.1
13 cotton bed sheet, c. 1960	white	good	6.3

[1] Specimens obtained from the University of Alberta
Historic Costume Collection, the Marian Center,
Edmonton, and from private donors.

Deacidifying Agents. Magnesium bicarbonate solution (pH
6.4) was made from laboratory grade magnesium hydroxide
using dry ice as the source of carbon dioxide (10).
Calcium hydroxide solution (pH 11.7) was prepared by
dissolving 4 g reagent grade calcium hydroxide in 2 L of
distilled water and stirring for two minutes. After
allowing the solution to settle for 24 hours at room
temperature, the clear solution was decanted. Methoxy
magnesium methyl carbonate (MMMC), available from Wei T'o
Associates, 21750 Main Street, Unit 27, P.O. Drawer 40,
Matteson IL 60443, was used in a dipping formula (solution
#2) and in a spraying formula (solution #12). Solution #2
contains trichlorotrifluoroethane as the primary solvent
and methanol as the secondary solvent for the MMMC.
Solution #12 contains ethanol as the primary solvent and
trichlorotrifluoroethane as the secondary solvent for the
MMMC and CO_2 as the propellant. Fabrics treated with

distilled water as a control and with magnesium
bicarbonate or calcium hydroxide were immersed for 30
minutes at room temperature, drained for 30 sec, then laid
flat to dry. Wei T'o solution #2 was applied by immersing
fabrics for 10 sec and draining for 10 sec. The Wei T'o
spray was applied to the fabrics as suggested in the Wei
T'o instruction leaflet. Further suggestions regarding
preparation and application of the finishes are provided
elsewhere (12). Not all fabrics in the study were treated
with calcium hydroxide or Wei T'o #12 spray.

Accelerated Aging. After treatment with deacidifying
agents, three naturally aged fabrics (#11-13 in Table I)
and the chlorine bleached fabric were exposed to
acclerated aging in the dark at 100% relative humidity
(RH), and 100°C for 15 days in a closed environment.
Aging under these conditions is reported to cause
hydrolysis of cellulose and produce primarily glucose and
xylose as short chain degradation products (12). The
aging chamber is described by Kerr et al. (3). Individual
specimens were rolled loosely around a glass rod, inserted
into a Diehls-Alder pressure test tube (Ace Glass, Inc.)
and suspended above one mL of distilled water in the
bottom of the tube. Tubes were capped with polyethylene-
lined bottle caps, and immersed in an oil bath for 15
days.

Test Methods. Color change after treatment and aging was
determined with a Hunterlab Tristimulus Colorimeter, Model
D25M-9. Six to ten specimens oriented in the same
direction were read individually and the color difference,
ΔE, determined in CIELAB units as detailed in AATCC Test
Method 153-1978 (13). In addition, a panel of experts in
color evaluation determined color change of specimens
using the Grey Scale for Staining. Aqueous extract pH of
fabrics was determined according to Tappi T509 os 77 (14)
using a Fisher Accumet pH meter. Moisture regain (%) of
10 specimens per fabric type was measured according to ASTM
Method D 2654-76 (15). In order to assess changes in
stiffness following treatment, the flexural rigidity of 10
specimens per fabric was determined using the cantilever
test described in ASTM Method D 1388-64 (15). Abrasion
resistance was evaluated according to ASTM Method D 3885-
80 (15). Twenty specimens per fabric were tested. The
number of cycles to rupture the specimen was recorded and
because so few cycles were required to rupture some of the
light-weight naturally aged fabrics, the numbers were not
rounded out as indicated in the test method. The
stainless steel bar on the Stoll Flex Abrader abraded the
specimens which were under a tension of 2.27 kg and
pressure of 0.45 kg. The appearance of fibers after
treatment with deacidifying agents and after abrasion was
evaluated with the aid of a scanning electron microscope.
Abrasion in this case consisted of 1000 cycles on the

Brush Pilling Tester with washed untreated test fabric mounted on the lower disc and treated fabric on the upper disc.

Statistical Analysis. When a sufficient number of observations was available for statistical analysis, a one way analysis of variance was carried out. The dependent variables, color change, flexural rigidity, moisture regain and flex abrasion, were measured against the independent variable, treatment. When a significant difference existed, Duncan's Multiple Range test ($\alpha = 0.05$) was used to establish which means formed a homogeneous subset, a group whose highest and lowest means do not differ by more than the shortest significant range for a subset of that size.

Results and Discussion

Color Change. Color difference and Grey Scale for Staining (GSS) values for new cotton, naturally aged, bleached and irradiated fabrics are presented in Table II. The naturally aged fabrics, although varying in extent of degradation, showed very little color change after treatment with the four deacidifying agents. Most had ΔE values of less than 3 CIELAB units and many treated with magnesium bicarbonate and calcium hydroxide solutions had ΔE values of less than 1 CIELAB unit, a color change which is barely detectable to the human eye. Most of the fabrics, when evaluated visually, were given GSS ratings of 4.0 or higher. These findings concur with Peacock's study (5) in which she found that none of the deacidifying agents visibly altered new fabrics. The Wei T'o dip and spray caused slightly more darkening of the naturally aged fabrics (#1-#13) than did magnesium bicarbonate. Although a statistical difference in color difference values existed in some cases, these differences in visual terms were minimal. The highly damaged cottons (#14-#16) showed a definite yellowing or darkening when treated. The Wei T'o treatments which are solvent-based rather than aqueous treatments caused slightly less darkening of these fabrics (ΔE values of about 5.3 to 6.0 CIELAB units) than did the magnesium bicarbonate or calcium hydroxide (ΔE values of 7.4 to 8.9). Since the 50 and 75 Mrad cottons were too fragile to be washed before treatment, they contained highly oxidized cellulose degradation products which could have reacted with the alkaline deacidification solutions and produced the yellow products. The overbleached cotton which is known to contain many carbonyl groups (7, 9) produced a distinct yellow color when immersed in calcium hydroxide solution, however, the color migrated into the bath and the fabric did not yellow noticeably.
    Three naturally aged cottons (#11-13) and the bleached new cotton (#14) were subjected to accelerated aging for 15 days at 100°C and 100% RH following deacidification. Color difference values are shown in Table III. The treatments

Table II. Color Difference (ΔE), Standard Deviation (SD) and Grey Scale for Staining Rating (GSS) of Naturally and Artificially Aged Cellulosic Fabrics Immediately After Deacidification[1,2,3]

Deacidifying Agent: Specimen	$Mg(HCO_3)_2$[2]			Wei T'o #2			Wei T'o #12			$Ca(OH)_2$[2]	
	ΔE	SD	GSS	ΔE	SD	GSS	ΔE	SD	GSS	ΔE	SD
0 new cotton	2.1	0.2	4.2[4]	1.2	0.2	4.5	0.9	0.3	4.8	–	–
1 linen cloth	0.3	0.1	4.7[4]	1.0	0.1	4.8	1.6	0.1	4.7	–	–
2 linen cloth	0.4	0.2	4.8	2.0	0.4	4.5	2.1	0.2	4.5	–	–
3 linen runner	1.3	0.9	4.3	2.3	1.1	4.3	1.5	0.5	4.2	–	–
4 cotton napkin	1.7	0.5	4.5	0.8	0.2	4.7	1.7	1.1	4.0	–	–
5 cotton shirt	1.2	0.9	4.5	2.1	1.4	4.0	2.8	1.1	3.8	–	–
6 cotton shirt	0.5	0.1	4.5[4]	1.0	0.1	4.8	0.9	0.1	4.7	–	–
7 cotton fabric	1.3	0.4	4.3	2.1	1.1	4.3	2.3	0.4	4.0	–	–
8 cotton gown	0.9	0.5	4.3	2.9	0.3	3.7	2.1	0.6	3.8	–	–
9 cotton sheet	0.7	0.1	4.8	1.9	0.3	4.5	2.1	0.5	4.5	–	–
10 cotton sheet	0.5	0.3	5.0	2.6	0.2	4.3	2.1	0.1	4.5	–	–
11 cotton sheet	0.4	0.1	–	0.9	0.1	–	–	–	–	0.5	0.2
12 cotton sheet	0.3	0.1	–	0.7	0.2	–	–	–	–	0.5	0.1
13 cotton sheet	0.3	0.1	–	1.1	0.1	–	–	–	–	0.5	0.1
14 cotton bleached	7.4	0.2	–	5.5	0.1	–	–	–	–	7.6	0.4
15 cotton 50 Mrad	8.3	3.3	3.0	5.7	0.8	3.2	5.3	1.0	3.2	–	–
16 cotton 75 Mrad	8.8	1.4	2.3	6.4	1.4	3.0	6.0	0.6	2.8	–	–

1 ΔE is calculated using L*, a*, b* values of the water-treated control as the initial values.
2 ΔE is the mean of 6 or 10 specimens; GSS, mean of 3 evaluations.
3 Verbal description of GSS ratings: 5 no change; 4 slight change; 3 noticeable change; 2 considerable change.
4 Expert panel described fabric as whiter than water-treated control.

Table III.   Color Difference (ΔE, CIELAB units) of
             Deacidified Cotton After Aging for 15 Days at
             100°C and 100% RH

	Treatment							
Specimen	Water		Mg(HCO3)2		Wei T'o #2		Ca(OH)2	
	ΔE	SD[2]	ΔE	SD	ΔE	SD	ΔE	SD
11 Cotton sheet	19[1]	2.0	14	2.3	17	3.6	13	2.1
12 Cotton sheet	15	0.4	14	0.8	17	0.7	16	0.5
13 Cotton sheet	21	2.5	18	0.8	19	1.0	18	2.0
14 Bleached new cotton	43	1.7	34	0.7	35	1.5	34	0.9

[1] Average of 10 specimens
[2] Standard deviation

significantly reduced the yellowing of fabrics #11, 13 and
14 compared to the water-treated control, although one
deacidifying agent was not markedly more effective than
the others.   The overbleached cotton became a deep tan
color during accelerated aging (ΔE = 43).   Nevel (7)
reports that the presence of aldehyde groups favors the
yellowing of cellulose during heating, particularly in the
presence of alkali and at temperatures above 170°C.   A
sweet smell was evident when the aging tubes were opened
and the water in the bottom of the tubes was brown
colored, suggesting the presence of water-soluble sugars.
The work of Erhardt and co-workers confirms that water
soluble sugars are formed (12).
     Conservators question whether deacidifying treatments
may be used on dyed cellulosic fabrics.   In Table IV, the
color difference values for seven dyed cotton fabrics
treated with Wei T'o #2 dip, #12 spray and magnesium
bicarbonate are shown.   Although the class of dye on the
cotton fabrics was not determined, the colors were not
sensitive to alkaline conditions.   Color difference values
were less than 1 CIELAB unit in 16 out of 21 cases, thus
the treatments caused no visually perceptible change in
dyed fabrics except for the fabric dyed yellow.   Daniels
(16) also found that magnesium bicarbonate and calcium
bicarbonate caused minimal color changes to organic
pigments on works of art on paper.   Because some dyes used
on cellulosic fabrics are sensitive to alkaline
conditions, it is essential that all colors in a historic
textile be tested before a deacidification treatment is
used.

Aqueous Extract pH.   The purpose of a deacidification
treatment is to neutralize internally-generated carboxyl
groups as well as acids from dyeing, finishing or exposure
to the environment.   Ideally an alkaline reserve should be
deposited in the fibers to combat future acidity.   Fabrics

Table IV.   Color Difference (ΔE, CIELAB units) of
            Commercially Dyed Cotton Fabrics After
            Deacidification[1],[2]

| | Treatment | | | | | |
| | Wei T'o #2 | | Wei T'o #12 | | Mg(HCO$_3$)$_2$ | |
Fabric Color	ΔE	SD[3]	ΔE	SD	ΔE	SD
yellow	2.3	0.1	2.3	0.3	1.1	0.2
green	0.7	0.3	0.7	0.1	1.0	0.3
mauve	1.1	0.3	0.8	0.2	0.8	0.1
coral	0.8	0.2	0.7	0.1	0.9	0.3
navy blue	0.7	0.2	0.7	0.3	0.7	0.4
light blue	0.7	0.3	0.4	0.1	0.8	0.3
aqua	1.2	0.2	0.6	0.2	0.7	0.1

[1] ΔE calculated using L*, a*, b* values of water-treated
control as initial values.
[2] Mean of 10 specimens.
[3] Standard deviation.

with aqueous extract pH values of 5.9 to 7.5 before
treatment (Table V), had pH values of about 9.5 after
treatment with calcium hydroxide solution and 10.2 to 10.5
after treatment with Wei T'o solution #2 and magnesium
bicarbonate.   These values are similar to those reported
previously by Kerr et al. (3), but higher than values
reported by Peacock (5).   All treatments were able to
maintain fabric pH values above 7.0 during accelerated
aging except for Wei T'o solution #2 on overbleached
cotton.   This cotton became very acidic during accelerated

Table V.   Aqueous Extract pH of Cellulosic Fabrics After
           Treatment and After Aging at 100°C and 100% RH
           for 15 Days [1],[2]

| | After Treatment[3] | | | | After Accelerated Aging | | | |
Specimen	H$_2$O	MB	WT#2	CH	H$_2$O	MB	WT#2	CH
11 old cotton	6.1	10.4	10.2	9.5	5.1	8.8	9.1	8.4
12 old cotton	6.1	10.3	10.3	9.5	5.3	8.3	9.3	8.5
13 old cotton	6.3	10.4	10.3	9.4	4.7	7.4	9.3	7.8
14 bleached cotton	5.9	10.2	10.3	9.4	3.8	7.5	6.6	7.8
0 new cotton	7.5	10.5	10.5	-	-	-	-	-

[1] pH of distilled water varied from 6.0 to 6.5.
[2] Mean of 2 determinations.
[3] Treatments: MB - Mg(HCO$_3$)$_2$; WT#2 - Wei T'o #2;
                  CH - Ca(OH)$_2$.

aging, probably because of the production of carboxyl groups from carbonyl groups generated during chlorine bleaching. A longer aging period is necessary to determine which deacidifying agent provides the best alkaline reserve.

Stiffness. The results of stiffness measurements, reported as flexural rigidity, are found in Table VI as well as weight gain (%) which occurred when specimens were treated with deacidifying agents or distilled water. Overall flexural rigidity is also reported as it facilitates comparisons among treatments. Differences in overall flexural rigidity, for example, indicate that the magnesium bicarbonate treatment caused the least change in stiffness (G_0 = 155 mg.cm) relative to the water-treated control (G_0 = 111 mg.cm). The Wei T'o #12 spray increased the stiffness about threefold whereas dipping specimens in Wei T'o #2 solution caused a tenfold increase in stiffness. In Figure 1c and 2d, fibers treated with Wei T'o #12 and #2, respectively, were once adhered together. The ruptured coating is visible between fibers. Fiber to fiber adhesion, no doubt restricts fiber mobility. This factor and perhaps changes in interfiber friction are likely responsible for the increase in stiffness (6). Peacock (5) reported similar increases in stiffness, but found that after subsequent aging, most samples had the same degree of stiffness. It was also noted in this study that slight manipulation of the stiffened samples made them flexible again. Fabrics dipped in Wei T'o solution #2 had the greatest weight gain (4.76%). Because flexural rigidity varies directly as w (the weight of a fabric) and as c^3 (the bending length) (15), increased interfiber bonding which stiffened the fabrics and increased bending length made a greater contribution to flexural rigidity than did weight gain.

Moisture Regain. Conservators have expressed concern that deacidifying agents on the surface of fibers will deliquesce and lead to moisture-related problems. This fear is unfounded. In Table VI, the moisture regain of new cotton treated with Wei T'o solution #2, #12, and magnesium bicarbonate is shown. The moisture regain of fibers increased a minimal amount, 0.2 to 0.4% above that of untreated cotton (8.2%).

Abrasion Resistance. Both new cotton fabric and four naturally aged cotton and linen fabrics were subjected to flexing and abrasion after treatment with deacidifying agents (Wei T'o #2, #12 spray and magnesium bicarbonate). The average number of cycles to rupture 20 specimens is shown in Table VII. The number of cycles to rupture fabrics 3, 8 and 10 was very small but this was not unexpected because the fabrics were made of very fine yarns. One fact apparent from the flex abrasion results

Table VI.   Mean Flexural Rigidity, Weight Gain and
            Moisture Regain of New Cotton Treated with
            Deacidifying Agents[1]

	Treatment			
	Wei T'o #2	Wei T'o #12	Mg(HCO3)2	Water
Flexural rigidity				
Warp (mg.cm)	1828±391[2]	505±71	266±16	196±17
Filling (mg.cm)	618±59	153±20	90±9	63±4
Overall flexural rigidity (mg.cm)	1063±152[3]	278±38	155±12	111±8
Weight gain (%)	4.76	2.01	3.89	–
Moisture regain (%)	8.63	8.65	8.41	8.19

[1] Flexural rigidity and moisture regain, mean of 10
specimens; weight gain, mean of 2 specimens.
[2] Mean ±95% confidence interval.
[3] Overall flexural rigidity, $G_o = (G_w G_f)^{0.5}$

is that new fabric behaved differently from old fabric;
its resistance to flexing and abrading decreased by a
factor of about 6 after deacidification although there is
no significant difference among treatments in cycles to
rupture the fabric. The flex abrasion resistance of
deacidified naturally aged fabrics, on the other hand, was
not significantly different from the water-treated
control. There are many sources of variability in flex
abrasion tests including changes in the abradant during
the test. During testing in order to avoid the build up
of a ridge of broken fiber fragments adjacent to the bar
over which the fabric was flexed, fiber debris was removed
every 30 cycles. It was noted that the ridge of abraded
fibers that formed when flexing and abrading the new
cotton did not occur when testing the naturally aged
fabric. Perhaps the aged fabrics broke so quickly due to
the fineness and brittleness of their yarns that fiber
debris did not have time to accumulate. Elongation,

Table VII. Flex Abrasion Resistance (Cycles to Rupture)
            of Deacidified New and Naturally Aged
            Cellulosics[1]

	Treatment			
Specimen	Water	Wei T'o #2	Wei T'o #12	Mg(HCO3)2
0 new cotton	891±31	143±6	154±6	160±9
3 old cotton	17±1	16±1	9±1	11±1
8 old cotton	22±3	18±2	15±2	18±2
9 old cotton	74±16	73±11	81±13	55±10
10 old cotton	21±8	33±8	23±8	23±5

[1] Mean±95% confidence interval, n=20.

elasticity and inter-fiber friction affect abrasion
resistance of a fabric (17). The immobilization of
fibers, visible in Figure 1, increased fabric stiffness
(flexural rigidity) and no doubt decreased abrasion
resistance of treated new cotton. New cotton dipped in
Wei T'o solution #2 had the highest flexural rigidity
(1063±152 mg.cm) and the lowest flex abrasion resistance
(143±6 cycles to rupture). The water treated control was
the most flexible (111±8 mg.cm) and had the highest
abrasion resistance (891±31 cycles to rupture). The
observation that treated fabrics do not stay stiff if
manipulated after deacidification is borne out by the flex
abrasion results. Although fabrics dipped in Wei T'o
solution #2 were initially about seven times stiffer than
the magnesium bicarbonate treated fabrics, their
performance when subjected to flexing and abrading (143±6
cycles to rupture) was not significantly different from
the magnesium bicarbonate treated fabric (160±9 cycles to
rupture). The flex abrasion test subjects a fabric to
thè magnesium bicarbonate treated fabric (160±9 cycles to
rupture). The flex abrasion test subjects a fabric to
flexing and rubbing over a steel bar while under tension
and pressure. These motions are sufficient to break
inter-fiber adhesion and soften a fabric. It should be
stressed that this test is more energetic than the
manipulation an historic textile receives if it is
carefully stored and displayed.

Scanning Electron Microscopy. One of the main reasons for
examining treated fabrics with the aid of a scanning
electron microscope was to determine whether the deacidification treatments left a crystalline deposit on the
surface of the fibers. In Figure 1 the water treated
cotton fibers look remarkably similar to those treated
with the Wei T'o solution and spray; fiber convolutions
and a wrinkled primary wall are readily visible (18). No
crystals can be seen. In Figure 1c there is evidence of a
coating which bound two fibers together but has now
separated. The magnesium bicarbonate treated fibers had a
distinctly mottled surface (Figure 1b), however, this did
not significantly affect abrasion resistance. Deacidified
fabrics were given gentle surface abrasion on the Brush
Pilling Tester, then observed under an electron microscope
to determine whether fiber damage had occurred. Abrasion
consisted of 1000 cycles on the Brush Pilling Tester
during which untreated new cotton rubbed the deacidified
new cotton in a circular motion. In Figure 2a and 2c,
water treated and Wei T'o sprayed fibers show splitting
parallel to fibrils, but it was not possible to determine
whether the splits were there before abrasion or resulted
from the surface rubbing. A crystal on the magnesium
bicarbonate treated specimen (Figure 2b) has not punctured
the fibers and the mottled surface appearance has
diminished. In Figure 2d, a coating which once held two

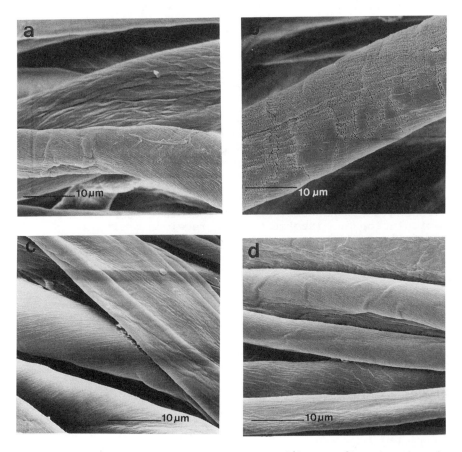

Figure 1. Appearance of new cotton fibers after treatment
with a) distilled water, b) magnesium bicarbonate,
c) Wei T'o #12 spray and d) Wei T'o #2 solution.

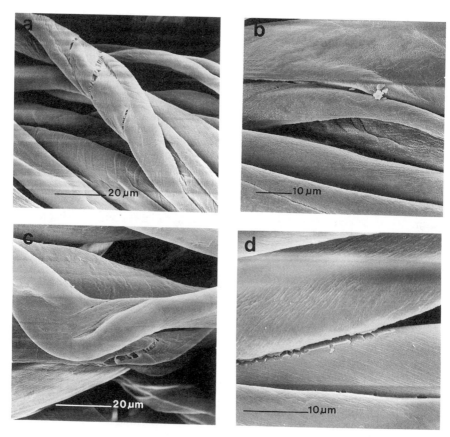

Figure 2. Effect of mild abrasion from a Brush Pilling Tester on new cotton fibers treated with a) distilled water, b) magnesium bicarbonate, c) Wei T'o #12 spray and d) Wei T'o #2 solution.

adjacent fibers together has broken. Cotton fibers after
mild abrasion often exhibit cracks, a smoother surface,
cuticle damage and broken fiber ends (18). Deacidified
samples showed some broken fibers and fibrillar splitting
but no more damage than the water treated control samples.
Large crystals capable of puncturing the cuticle were not
found.

Conclusions

This study provides some answers to questions about
physical changes which occur when a textile is treated
with four deacidifying agents, magnesium bicarbonate,
calcium hydroxide, and Wei T'o solution #2 and #12 spray.
Magnesium bicarbonate forms a slightly acidic solution
when freshly prepared (10), however, once a fabric is
treated and exposed to the air, alkaline magnesium
carbonate forms on the fabric. If a fabric is treated
with this solution or the highly alkaline calcium
hydroxide solution (pH 11.7), all colors on dyed or
printed textiles should be tested for sensitivity to
alkalies. The Wei T'o products, solution #2 and #12
spray, are applied from a nonaqueous base (fluorocarbon
plus ethanol) and, as such, are more suitable than
magnesium bicarbonate or calcium hydroxide for textiles
which are adversely affected by water. The Wei T'o spray
is particularly suitable for an artifact that is too large
or cumbersome to immerse, although practise applying the
spray evenly is essential.
    Historic cellulosic textiles which are not highly
oxidized should have negligible color change upon
treatment and during subsequent aging, should retain their
whiteness better than untreated fabric. Deacidification
is not recommended for highly degraded cellulosics which
are yellowed from oxidation unless they can be washed
before treatment. It is important to remove oxidized
water-soluble degradation products so that they can not
react with the alkaline treatments and cause further
yellowing. Each agent in this study deposited an alkaline
reserve which was capable of maintaining the pH of old
fabrics above 6.6 during accelerated aging at 100°C and
100% relative humidity.
    One major physical change which occurs when fabrics are
treated, particularly with Wei T'o solution #2, is an
increase in stiffness. Since stiffness affects fabric
drape, a distinctive characteristic of a textile, a change
in this characteristic is not acceptable except perhaps in
a textile which will be mounted flat. Scanning electron
photomicrographs show that some interfiber adhesion occurs
during treatment with this agent and #12 spray. The
stiffness disappears quickly if the fabric is manipulated
gently. In old fabrics which were treated with these
agents, the abrasion resistance was no different than the
water treated control. The abrasion resistance of new

cotton decreases considerably after deacidification,
however, and this finding suggests that conservators
should use caution in treating new fabrics if they are to
be used in an application where the fabric is subjected to
abrasive forces or it must have a soft drape. The results
of this study should assist a conservator who seeks
information about the physical changes which may occur
when a cellulosic fabric is treated with four common
deacidifying agents.

Acknowledgements

The authors wish to acknowledge the financial assistance
of the Central Research Fund at the University of Alberta,
Edmonton, Alberta. George Braybrook, head of the SEM
facility in the Department of Entomology, University of
Alberta, provided valuable assistance with the SEM photomicrographs.

Literature Cited

1. Block, I. In Preprints: IIC Washington Congress;
   Bromelle, N.S.; Thomson, G., Eds.; Intern. Inst. for
   Conserv. of Historic and Artistic Works: London,
   1982, pp 96-99.
2. Kerr, N., Hersh, S.P., Tucker, P.A., Berry. G.M. In
   Durability of Macromolecular Materials; Eby, R.K.,
   Ed.; ACS Symposium Series No. 95; American Chemical
   Society: New York, 1979; pp 357-369.
3. Kerr, N., Hersh, S.P., Tucker, P.A. In Preprints:
   IIC Washington Congress; Bromelle, N.S.; Thomson, G.,
   Eds.; Intern. Inst. for Conserv. of Historic and
   Artistic Works: London, 1982, pp 100-103.
4. Kerr, N., Hersh, S.P., Tucker, P.A. In Preprints:
   ICOM Comm. for Conserv. 7th Triennial Meeting; de
   Froment, D., Ed.; Intern. Council of Museums: Paris,
   1984, pp 9.46-50.
5. Peacock, E. Studies in Conserv., 1982, 28, 8-14.
6. Grosberg, P. In Surface Characteristics of Fibers and
   Textiles, Part II; Schick, M.J., Ed.; Dekker:
   New York, 1977; pp 563-575.
7. Nevell, T.P. In Cellulose Chemistry and Its
   Applications; Nevell, T.P.; Zeronian, S.H., Eds.;
   Horwood: Chichester, 1985, pp 223-256.
8. Berry, G.M., Hersh, S.P., Tucker, P.A., Walsh, W.K.
   In Preservation of Paper and Textiles of Historic and
   Artistic Value; Williams, J.C., Ed.; Advances in
   Chemistry Series No. 164; American Chemical Society:
   New York, 1977; pp 228-284.
9. Burgess, H.D., Hanlan, J.F. J. of IIC-CG 1, 2, 1980;
   15-22.

10. Wilson, W.K., McKeil, M.C., Gear, J.L., MacLaren, R.H. _American Archivist_, 1978; 41, 67-70.
11. Jennings, T., M.Sc. Thesis, University of Alberta, Edmonton, AB, 1985; pp 77-80.
12. Erhardt, D., Von Endt, D., Hopwood, W. In _Preprints of the 15th AIC Meeting_; Brown, A.G., Ed.; Am. Inst. for Conserv.: Washington, 1987, pp 43-45.
13. _AATCC Technical Manual_, Vol. 58, AATCC: Research Triangle Park, NC, 1982/83.
14. _TAPPI Standards and Provisional Methods_, Tech. Assoc. Pulp Paper Industry: Atlanta, GA, 1977.
15. _Annual Book of ASTM Standards_, Vol. 07.01, Am. Assoc. for Testing and Materials: Philadelphia, 1983.
16. Daniels, V. In _Preprints: IIC Washington Congress_; Bromelle, N.S.; Thomson, G., Eds.; Intern. Inst. for Conser. of Historic and Artistic Works: London, 1982; pp 66-70.
17. Galbraith, R.L. In _Surface Characteristics of Fibers and Textiles_, Part 1, Schick, M.J., Ed.; Dekker: New York, 1975; pp 193-204.
18. de Gruy, I.V., Carra, J.H., Goynes, W.R. _The Fine Structure of Cotton, An Atlas of Cotton Microscopy_; Dekker: New York, 1973; p 198.

RECEIVED March 27, 1989

Chapter 11

Heat-Induced Aging of Linen

Howard L. Needles and Kimberly Claudia J. Nowak

Division of Textiles and Clothing, University of California, Davis, CA 95616

Linen made from flax fibers was aged by heating at
180°C in air for periods of up to ten hours. The
heat-treated linen became progressively darker and
more red and yellow (brown) in color and showed
progressive losses in tensile and abrasion
properties. Wide angle X-ray scattering suggested
that the heat-aged linen was somewhat less
crystalline than untreated linen. Scanning electron
microscopy showed that heat aging caused long
crevices longitudinal to the fiber axis and passing
through nodes in the flax increasing the accessibility and surface area of the fibers. Heat-aged
linen dyed to lighter and slightly different shades
and had fewer dyeing sites available for direct dyes
than untreated linen. Therefore, flax oxidation
during heating apparently led to some breakdown of
crystalline regions in the cellulose, but did not
provide additional dyeing sites. The loss in dyeing
sites is thought to be due to heat-induced
crosslinking of the amorphous regions in the flax.

Linen textiles made from flax fibers have been known and used by
mankind since antiquity (1). Flax has been used in many textile
constructions including fine linen fabrics, laces, embroideries, and
bridal fashions, and many historic linen textiles have become part
of permanent museum collections. Older linen fabrics and laces are
prized for their natural creamy color and luster and often have been
recycled and reused. However, little is known about natural aging
of linen. Most aging studies for cellulosics such as linen have
involved accelerated heat-induced aging.

Kleinert (2) observed that ancient linens exhibited low degrees
of polymerization, loss of strength, severe fiber deterioration,
high overall crystallinity but short crystallites and a high degree
of oxidation. Hackney and Hedley (3) examined the aging of linen

0097–6156/89/0410–0159$06.00/0
© 1989 American Chemical Society

canvas kept in the Tate Gallery under various conditions for 24 years. They found that light exposure during this period caused significant changes in the linen including losses in tensile strength and slight discoloration. Sulfur dioxide present in the air was shown to contribute to this deterioration in properties. Continuous changes in temperature and humidity during the exposure of the linen were also implicated. Peacock (4) studied moist heat aging of linen at 70°C and 50% RH for 21 days. The heat-aged linen samples lost weight, discolored to a light grey brown, became more flexible and lost tensile strength. Deterioration and discoloration of other fabrics made from cellulosic fibers can be accelerated by heating at elevated temperatures (5-13). These studies have shown that heat aging of cellulosics causes accelerated discoloration, loss in tensile properties, increased fiber breakage, reduction in the degree of polymerization, reduction in the degree of swelling, in some instances increases in crystallinity, increases in the carbonyl and carboxyl group content, decreases in moisture content, and reduced the dye uptake of the fiber. These property changes are similar to changes in natural-aged linens.

Since so little is known about heat aging of linen, we examined the dry heat-induced aging of linen at 180°C from 1 hr. to 10 hrs. We have studied the effect of heating on the color, on the dry and wet tensile properties, on the abrasion characteristics, and on the dyeing and resultant color properties of the linen. We also examined the effect of heat treatment on the crystallinity of flax by wide angle X-ray scattering (WAXS) and on the surface morphology of flax fibers by scanning electron microscopy (SEM).

Experimental

Materials. The linen was a bleached handkerchief linen (#L-61) with a thread count of 60 x 50 picks/inch from Testfabrics, Inc. All direct dyes were commercial grade from Aldrich Chemical Co., while sodium sulfate was reagent grade from Mallinckrodt Chemical Co.

Heat-Induced Aging. Linen samples (12 x 4") were washed in deionized water containing 0.1% Triton X-100 surfactant, rinsed, and air dried. The samples were aged at 180°C in a forced draft laboratory oven for 1, 3, 5 or 10 hrs. The samples were removed from the oven and conditioned at 21°C and 65% RH prior to testing.

Test Methods. Color differences were measured on a MacBeth MS 2000 Color Spectrophotometer using the CIELAB color system and the yellowness index (YI). Three color readings were made and averaged for each sample. Color changes are the differences between heat-aged and untreated linen samples and are expressed as ΔL^*, Δa^*, Δb^*, ΔE, and ΔYI color differences (Table I). Dry and wet tensile properties of yarns (measurement of fifty yarns for each sample) were measured using an Instron textile tester using ASTM Method D2256 and a three inch gauge length. The relative changes in tensile properties for the heat aged samples compared to untreated linen are given in Table II. Multidirectional abrasion tests (five specimens for each sample) were carried out on a Universal Wear

Table I. Heat-Induced Color Changes

Heating Time (hr.)	ΔL*	Δa*	Δb*	ΔE	ΔYI
1	-6.0	-1.6	20.4	21.3	20.9
3	-8.8	-0.7	24.7	26.3	34.0
5	-9.5	-0.1	25.7	27.4	34.3
10	-16.4	1.9	31.6	35.7	47.8
Control L* = 94.8		a* = 1.75	b* = -7.1		

Table II. Relative Changes in Tensile and Abrasion
Characteristics of Heat-Aged Linen

Heating Time (hr.)	Breaking Strength		Elongation at Break		Energy to Break		Abrasion
	Dry	Wet	Dry	Wet	Dry	Wet	Dry
1	0.75	0.64	0.94	0.62	0.65	0.50	0.77
3	0.54	0.36	0.76	0.58	0.38	0.21	0.71
5	0.49	0.31	0.82	0.66	0.35	0.18	0.58
10	0.39	0.17	0.41	0.54	0.27	0.07	0.41

Abrasion Tester from Custom Scientific Instruments using ASTM Method
D3886 (Table II). Scanning electron microscopy (SEM) was performed
on gold-coated samples on a ISI DS-130 Scanning Electron Microscope
using a Lab 6 filament at 10 kV (Figure 1). Wide angle X-ray
scattering (WAXS) of untreated and heat-aged linen was carried out
on a DIAN-XRD 800 diffractometer giving 50 kV Cuk$_\alpha$ radiation at 15
mA with scanning performed from 8 to 30° at 1.6 deg/min (Figure 2).

Dyeing Procedure. Untreated and heat-treated 1 g samples were dyed
from infinite dyebaths containing Direct Red 2, Direct Violet 51, or
Direct Blue 7 (Figure 3) and 20% (owf) sodium sulfate. The liquor
ratio for each dyeing was 100:1, and the dyeings were carried out
for 1 hr. at 100°C. After dyeing, the samples were thoroughly
rinsed with hot water, followed by deionized water, and allowed to
air dry prior to measurement of color differences between untreated
and heat-aged samples by the method described above (Tables III, IV,
V).

Results and Discussion

Color and Tensile Property Changes in Heat-Aged Linen. On heating
at 180°C from 1 to 10 hrs, the linen fabric became progressively
darker (-ΔL*), slightly more red (+Δa*) and progressively more
yellow (+Δb*) in character. The overall color difference (ΔE) and
yellowness index (ΔYI) progressively increased with heating time
(Table I). The combined effect was a progressive darkening and
browning of the linen. Such heat-induced color changes have been
observed before for linen (4) and cotton (13), at lower temperatures
or shorter heating times, but full characterization of the color

a

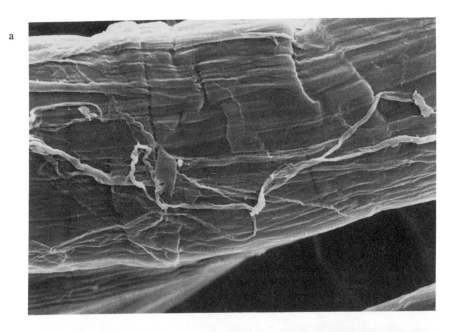

b

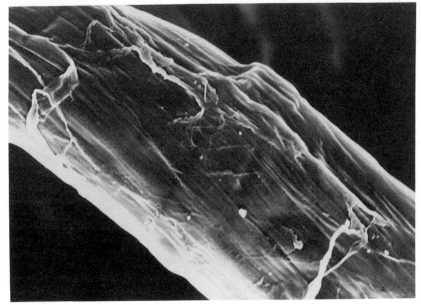

Figure 1. Untreated linen: a, × 5160; and b, × 6930. *Continued on next page.*

c

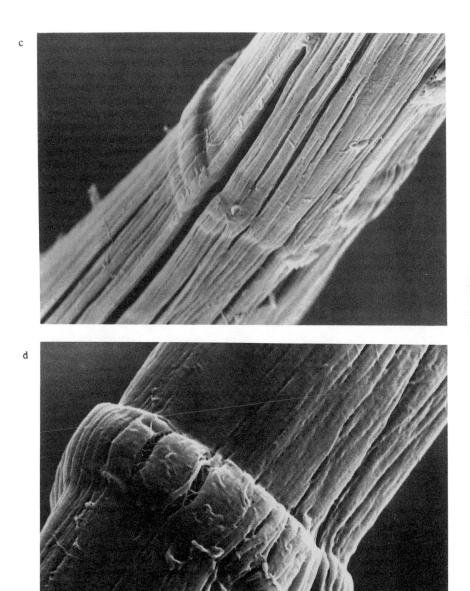

d

Figure 1. *Continued.* Linen heat-treated 10 h: c, × 6990; and d, × 9570.

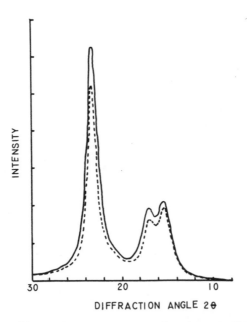

Figure 2.   Wide angle X-ray scattering (WAXS) diffractograms
_____ - untreated linen, — — — — - linen heat-
treated for 10 hrs.

Direct Red 2

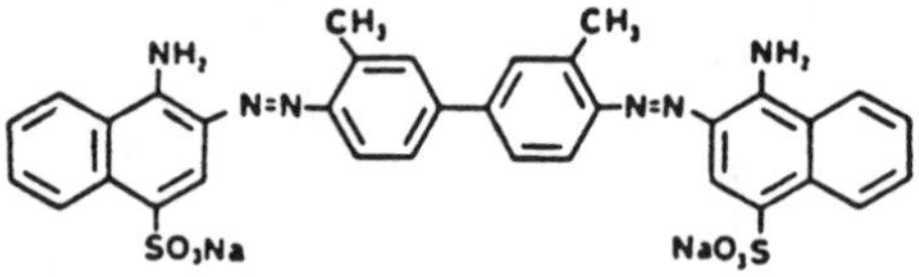

Direct Violet 51

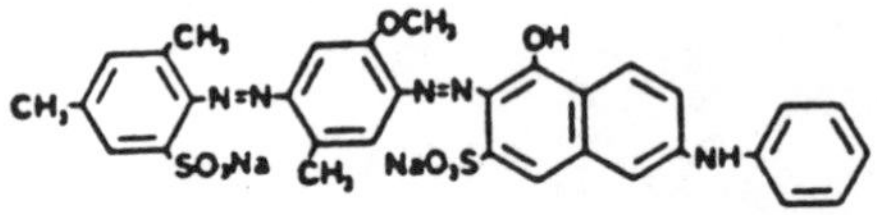

Direct Blue 71

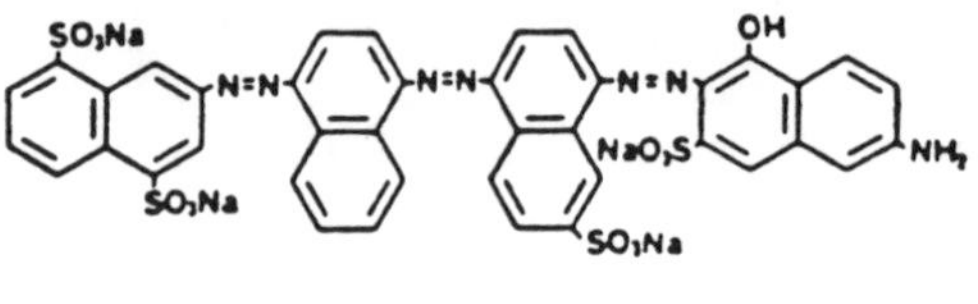

Figure 3.   Direct dye structures

changes that occurred has not been made. The browning of the fabric represents oxidation of the linen to form conjugated unsaturated structures that absorb light in the violet and blue regions of the visible spectra.

Because of the high temperature (180°C) and extended heating time involved (up to 10 hrs.) in this study, it was expected that the linen would undergo extensive degradation. This is reflected in the extensive loss in tensile and abrasion properties of the linen with wet tensile properties being much more severely affected than dry tensile properties (Table II). Even after 1 hr. of heating, a significant drop in tensile properties was found with wet energy to break values being the most sensitive-indicator of tensile property deterioration. Although the heat-treated linen is progressively more readily abraded, the change in abrasion characteristics seems to be a less sensitive indicator of linen deterioration. In summary, as the heating time increased, linen progressively decreased in breaking strength, broke at shorter elongations, and exhibited lower energy at break and fewer cycles to abrasive failure due to degradation. Wet tensile properties were much more affected than dry tensile properties. Losses in the tensile strength of heat-aged linen (4) and cotton (5,10,13) were observed previously, but generally not at these levels of degradation with the exception of the work of Berry, Hersh, Tucker, and Walsh (10). Back and coworkers (6,7) using lower temperatures of heating have observed an increase in the wet tensile strength of cellulose, a finding contrary to our study presented here for high temperature aging of linen.

Scanning Electron Microscopy and Wide Angle X-Ray Scattering. Heat-aged (10 hr.) and untreated linen samples were examined by scanning electron microscopy (SEM) (Figure 1) and wide angle X-ray scattering (WAXS) (Figure 2). Scanning electron microscopy showed that flax fibers from the untreated linen were characteristic of flax with periodic nodes and subtle fibrillar structure running longitudinally along the fiber axis (Figure 1a, 1b). Flax fibers from heat-treated (10 hr.) linen have similar surface morphology compared to untreated linen (Figure 1c, 1d), but also have long and deep periodic cracks running parallel to the fiber axis. These heat-induced cracks might be expected to increase the accessibility of the flax fibers to dyes.

WAXS diffractograms of untreated and 10 hr. heat-aged linen were the same except that the intensity of the diffractogram of heat-aged linen was less than that for untreated linen. This suggests that heat aging does not particularly affect the size and shape of crystallites in the linen, but that heating slightly reduced the total crystallinity and slightly increased the amorphous areas contained in the linen. Segal and coworkers (14) have shown a similar drop in intensity of x-ray diffractograms as cotton was partially decrystallized. This is contrary to the findings of other workers using other techniques that suggest that heating increased the crystallinity, decreased the degree of polymerization (DP), and introduced crosslinks into the cellulose (9,10,12).

Dyeing Properties of Heat-Aged Linen. The dyeing and resultant color properties of heat-aged linen were compared to these properties for untreated linen using three direct dyes of differing structures (Figure 3 and Tables III, IV, V). Heat-treated and untreated linen dyed to medium shades with these direct dyes. However, the heat-treated linen samples dyed to progressively lighter (ΔL*) and slightly different shades (Δa*, Δb*) than found for untreated linen. Perceptible color differences were observed (ΔE). Dyeing at these depths of shade effectively covered heat-induced darkening (ΔL*) and browning (Δa*, Δb*) of the linen. Other workers have observed that heat-aged cellulosics dye to lighter shades than untreated cellulosics (5,8,13). This reduction in dye adsorption by the heat-aged cellulosics has been attributed to heat-induced changes in the accessibility of the amorphous regions through processes such as crosslinking or through actual loss of some of the amorphous regions through crystallization.

Table III. Color Changes in Heat-Aged Linen
Dyed with Direct Red 2

Heating Time (hr.)	ΔL*	Δa*	Δb*	ΔE
1	1.9	1.0	0.1	2.1
3	2.4	0.7	0.1	2.5
5	3.6	0.9	-0.3	3.7
10	4.4	0.6	-0.1	3.5
Dyed Control	L* = 35.9	a* = 48.6	b* = 28.2	

Table IV. Color Changes in Heat-Aged Linen
Dyed with Direct Violet 51

Heating Time (hr.)	ΔL*	Δa*	Δb*	ΔE
1	0.4	0.7	-0.5	0.9
3	1.5	0.6	-0.1	1.6
5	0.8	0.7	-0.2	1.1
10	1.5	1.3	-0.8	2.1
Dyed Control	L* = 20.4	a* = 9.1	b* = -8.8	

Table V. Color Changes in Heat-Aged Linen
Dyed with Direct Blue 71

Heating Time (hr.)	ΔL*	Δa*	Δb*	ΔE
1	1.6	0.3	-1.4	2.1
3	2.8	0.6	-2.4	3.7
5	3.3	0.6	-2.8	4.4
10	5.0	0.7	-3.5	6.1
Dyed Control	L* = 21.9	a* = 0.1	b* = -9.4	

As heating time increased, the linens dyed with Direct Red 2 to progressively lighter shades (+ΔL*) and were slightly more red (+Δa*) in character compared to untreated linen (Table III). The heat-aged linens dyed with Direct Violet 51 to progressively lighter (+ΔL*) and slightly more red (+Δa*) and blue (-Δb*) shades compared to untreated linen. Finally heat-treated linen samples dyed with Direct Blue 71 to progressively lighter (+ΔL*) and red (+Δa*) and blue (-Δb*) shades than untreated linen. Samples dyed with Direct Blue 71 exhibited the greatest color differences (ΔE) suggesting that heat-aging lowered the number or accessibility of dyeing sites most for this larger triazo direct dye. These findings are consistent with the proposal that although heat aging causes a slight increase in the amorphous content of the flax, crosslinking also occurs lowering the accessibility of these dyes.

Conclusions

As expected, heat-aging caused the linen to darken and brown and to undergo a reduction in tensile and abrasion properties. Heat aging led to formation of long deep crevices longitudinally along the fiber axis and passing through the nodes in the flax. Although these crevices appear to increase the accessibility and surface area of the fibers, dye uptake did not increase. Heat-aged linen was less dyeable than untreated linen with the largest of the three dyes having less access to dye sites within the heat-aged linen. Although wide angle x-ray diffraction suggests that there is an apparent net increase in the amorphous regions present within the heat-aged cellulose, there is a net decrease in dyeing sites available. These findings suggest that the heat-induced crosslinking within the amorphous regions of the flax occurs lowering the number of accessible dyeing sites.

Literature Cited

1.   Cook, J. G. Handbook of Textile Fibres. I. Natural Fibres; Merrow Publishing Co., Ltd., 1984; p 4.
2.   Kleinart, T. N. Holzforschung 1977, 21, 77.
3.   Hackney, S.; Hedley, G. Stud. Conserv. 1981, 26, 1.
4.   Peacock, E. E. Stud. Conserv. 1983, 28, 8.
5.   Hessler, L. E.; Workman, H. Textile Res. J. 1959, 26, 487.
6.   Back, F. L.; Klinga, L. O. Svensk Papperstiden. 1965, 66, 745.
7.   Back, F. L.; Didriksen, E. I. Svensk Papperstiden. 1969, 72, 687.
8.   Rushnak, I.; Tanczos, I. Preprint, IUPAC Symposium on Macromolecules; Helsinki, 1972; V, 127.
9.   Zeronian, S. H. In Cellulose Chemistry and Technology; Arthur, Jr., J. C., Ed., ACS Symposium Series; 1977; 48, 189.
10. Berry, G. M.; Hersh, S. P.; Tucker, P. A.; Walsh, W. K. In Preservation of Paper and Textiles of Historic and Artistic Value; Williams, J. C., Ed., Advances in Chemistry Series; 1977, 164, 228.
11. Shafizadeh, F.; Bradbury, A. G. W. J. Appl. Polym. Sci. 1979, 23, 1431.
12. Lewin, M.; Guttmann, H.; Derfler, D. J. Appl. Polym. Sci. 1982, 27, 3199.
13. Brushwood, D. E. Textile Res. J. 1988, 58, 309.
14. Segal, L., Creely, J. J., Martin, Jr., A. E.; Conrad, C. M. Textile Res. J. 1959, 19, 786.

RECEIVED January 24, 1989

Chapter 12

Treatment of Tapa Cloth with Special Reference to the Use of the Vacuum Suction Table

Sara Wolf Green

Textile Museum, 2320 S Street, NW, Washington, DC 20008

The treatment of ethnographic artifacts like
tapa cloth require an understanding of the
material and its manufacture. The literature
of the conservation of tapa is reviewed, and
new techniques for filling voids are presented. These techniques include the use of
hand-cast fills, and fills cast on the
vacuum suction table. Case studies from 1980,
1986 and 1987 are summarized to illustrate the
evolution of treatment techniques for tapa.

Over the past three years, the treatment of tapa cloth
has re-emerged as a topic of interest to conservators of
ethnographic materials, as evidenced by papers presented
at meetings of the AIC (1), CCI Symposium '86 (2,3), and
the Jubilee Conference at the University of London (4).
While a number of questions remain as to the best way to
stabilize this material, certain techniques have been
proposed which have shown promise, including the use of
cast fills for filling voids.

Cloth made from the inner bark (secondary phloem) of
various plants is found throughout many of the tropical
regions of the world. Although the origins of its
production are unknown, they have been chronicled as
early as the 6th-century B.C. in China (5). The bark
cloth of the Pacific, more commonly referred to by the
Polynesian word "tapa", has been produced from the inner
bark of various species of trees of the genera
Broussonetia R. (paper-mulberry), Artocarpus
(breadfruit), and Ficus (fig species) (6). While
examples of each of these types are well documented, the

0097–6156/89/0410–0168$06.00/0

finest quality tapa is that produced from the
Broussonetia papyrifera.
   While there are certain similarities between tapas
produced world-wide, the investigation of treatment
techniques described here have been applied only to
examples from Fiji and Samoa.   There are manufacturing
differences between these two island groups, primarily in
the methods by which decorations are most commonly
applied.   However, the cloths themselves are similar in
thickness, texture, and stability of the pigments, and
the treatment techniques applied were found to be
interchangeable.

Cultivation and Harvesting.   The cultivation and
production of high quality bark cloth is not accidental.
The trees are planted in specific areas chosen for the
quality of soil, good drainage, and adequate rainfall.
During the growing seasons, side branches are carefully
pruned to prevent holes from forming in the bark, and
harvesting is carefully calculated so that the sapling is
mature enough to produce a substantial bark, but not so
old that the bast fibers are coarse and difficult to
strip.
   In many areas, both men and women are involved in the
production of tapa, and while specific duties are
generally the province of either men or women, the
assignment of tasks vary by area even within a culture.
For example, in some areas of Fiji (Navatusila district,
Viti Levu), men cut down the trees and the women take
over the production from that point (7).   In other areas
(Lau Islands), men may also participate in the
acquisition of pigments and dyes (8).
   The bark is stripped from the trees by making a cut
through the outer bark layers, and then pulling the bark
off in long strips.   The inner and outer bark layers are
stripped from each other and the inner bark is then
rolled to prevent curling.   The inner bark is then
usually soaked in water to loosen remaining bits of outer
bark and keep the inner bark supple.   In Fiji, the
remaining outer bark is removed by scraping with shells
(or more recently with a blunt cane knife), the inner
bark folded, and the water pressed out.   At this point,
the bark (solely secondary phloem, composed of sieve and
parenchyma cells and fibers) can either be worked into a
sheet of tapa cloth, or dried and stored for later use.

Manufacture.   The basic tools for tapa production are a
hardwood anvil and beater.   Two to four layers of the
bark are laid out on the anvil which has a smooth and
slightly convex surface.   Heavy, widely grooved mallets
are used at the beginning of the beating process to
spread and felt the fibers.   More finely grooved, lighter

mallets are used at a later stage to smooth the ribbed
texture of the surface, although elaborately carved
beaters may be used instead for the specific purpose of
leaving imprints on the tapa cloth (9).
    Larger tapas are made by seaming together several
sheets of the cloth. Sewing has also been used in the
construction of tapa articles, notably in Africa, South
America and Hawaii. In Fiji, the finest cloths were
joined by splitting one piece between the layers and
inserting the edge of a second tapa into the split. The
seam was then felted by additional beating to become an
almost imperceptible join. Holes in the cloth were
filled by adding a scrap of bark or pulling the edges of
the hole together and refelting. It is more common to
see pasted joints and fills in contemporary cloths in
Fiji. Samoan cloths on the other hand, have
traditionally been lengthened and mended by pasting.
Starch from the tubers of arrowroot (10) or cassava (11)
continue to be used for that purpose.

Decoration
    While some tapas are left undecorated, or simply smoked
to produce a warm brown color, most are decorated with
painting, rubbed designs produced by placing a raised
relief under the cloth and rubbing the surface with dues,
"printing" with leaves and ferns dipped in dye or paint,
or in Fiji, by stenciling.
    Sources for coloring tapas vary by area. Turmeric is
one of the most widely distributed sources for yellow in
the Pacific (12). Black and browns come from a variety
of sources including soot from the burning of candlenut
and sap from the mangrove. The reds of Tahiti were
produced from a variety of berry juices, but their colors
are often fugitive and may have faded to brown. There
are also red clays in several of the island groups which
are ground and mixed, for example with mangrove sap. The
tapas of New Guinea feature the largest variety of
pigment decorations from clays, soapstone, charcoal and
burnt lime (13).

Use. The uses to which tapa has been put are as diverse
as the cultures that produce it. It has accompanied
ceremony and has been used for everyday apparel, wall
coverings, tribute, mosquito netting and bed coverings,
to mention a few. Tapa has even been used for the
printing of commemorative newspapers in the South
Pacific.
    Another important use for tapa has been as waterproofed
coverings, or "rain-wear". In polynesia, tree pitch, and
to a lesser extent, coconut oil, have been used to render
tapa water impermeable. This stiff, often brittle tapa
produces preservation difficulties quite different from

painted or dyed tapas. Coconut oil or pitch coatings can
also be found on painted tapas where it is used as a
"glaze" or sometimes as a pigment binder over specific
areas of design. This application of oil or pitch also
serves to render the tapa stiff in the areas to which it
is applied.

Museum collections often contain tapas which are
extremely large in size, but are still only a portion of
a much larger tapa. Gift or ceremonial tapas could reach
200 feet or more in length. These were often cut apart
and distributed among family members, but even these
smaller pieces could extend beyond 25 feet. These
extremely large tapas present not only difficulties of
interpretation in an exhibit because they have been
completely removed from context, but also present obvious
problems for the mechanics of display and storage.

Condition

Like any other kind of collection, the condition of tapa
is highly dependent upon the quality of manufacture, the
extent to which it has been used, the manner in which it
has been used, the quality of storage, display methods,
and the environment.

Tapas have historically been folded for storage, a
factor which accounts for a great deal of deformation and
breakage of the fibers. In Fiji, the house rafters were
the typical storage location for folded and tied bundles
of tapa. Dirt embedded in these cloths may contain a
greasy, sooty accumulation from cooking fires if they
were collected before the turn of the century. Since
cooking fires were moved out of house interiors under
British missionary influence about that time, a sooty
accumulation on a tapa from that area may provide a clue
as to its age.

Other typical problems associated with tapa and its
deterioration include delaminations (separations between
the felted layers), cracks, holes (punctures, losses and
insect damage), and tears. Tapas are frequently acidic
and consequently brittle. The range of pH measured on
five tapas tested during a 1978 treatment project were
between 4.0 and 5.6 (measurements on recently
manufactured, contemporary unprinted tapa were slightly
acidic, ranging from 6.4 to 6.9). In addition to soot,
dirt and staining may be present from any number of
sources including soils from wear and oil used as bodily
decoration in ceremonies.

The thin smoked Fijian tapa which was used for turbans
is generally so fragile that few examples remain. Those
pieces which are extant are frequently so brittle that
they cannot be handled without pieces breaking off.

Whether these pieces can or should be treated is a
serious question.

Treatment

The treatment techniques for tapa fall into the following
categories: (a) cleaning, (b) crease removal, (c) tear
repair and/or backing, and (d) filling voids.

Cleaning. Careful vacuuming is an acceptable method of
removing surface dust and dirt from a tapa. Beyond that,
non-solvent drycleaning and wet cleaning have been
proposed for a variety of reasons, including the desire
to remove surface dust accumulation, grime and chemical
pollutants, and excess acidity (14). As with many other
ethnographic materials, it is possible that the soils
present may represent valuable information on history and
use. If so, surface accumulations should probably be
retained. If it can be determined that ethnographic
evidence will be retained in the cleaning process, or if
soils do not represent important cultural information,
several alternatives are available.
   The use of vinyl eraser crumbs may be viable for the
removal of some surface soils. However, the amount of
pressure used and the direction the eraser crumbs are
rolled are critical. Moving across or against the
"grain" of the tapa will cause bits of fiber to be picked
up on the eraser crumbs.
   Water treatments can significantly improve the
flexibility of tapa. However, some of the pigments and
dyes used to decorate tapas are fugitive. In tests on
more than 30 Fijian and Samoan tapas, the decoration or
painting of only 4 (tapas used as newspapers and printed
with ink) were completely stable to immersion. Natalie
Firnhaber (Smithsonian Institution, personal
communication, 1986) reported few difficulties with water
treatments for Hawaiian tapa and considers washing to be
a viable option. She has also had success with the use
of dampened blotters to remove excess acidity.
   While the bleeding of pigments and dyes may be more
typical for Fijian and Samoan pieces, the potential for
individual eccentricities in tapas from other island
groups should not be underestimated. In the tests on the
30 Fijian and Samoan tapas mentioned above, movement of
pigments was observed within 10 to 15 minutes of the
application of dampened blotters.
   Even if pigments are completely stable to wet
treatments, there is the potential for starch adhesive
joins and fills adhered with starch to give way. In
addition, delaminations between layers may also occur, or
increase if already present. It has also been observed
that tapa may decrease in strength while wet, and can be
stretched and distorted if not carefully supported.

The drying of a damp or wet tapa also requires special
handling. During the manufacturing process, tapas are
sometimes held under tension by placing rocks around the
edges as they dry. They may also be smoothed by hand
while still damp, or the artist may rely on the beating
procedure to flatten the sheets. Because of the tendency
to retain the "fiber memory" of the curved shape of the
bark, a tapa will tend to warp and curl with changes of
relative humidity, and when it is wetted. As a
consequence, a tapa will need to be weighted along the
edges after a water treatment so that it will dry as flat
as possible. In general, overall weights are not
recommended during the drying process because a tapa has
an irregular enough surface that an overall weighting may
introduce creases.

Wet cleaning with detergents (15), as well as
deacidification, or buffering with alkaline compounds
(case study summary for Document 55, Fiji Museum) have
also been suggested as options for the treatment of tapa.
These two suggestions show clearly the amount of overlap
between routine textile and paper treatments that have
been applied to tapa. In both cases, however, these
techniques have been applied without the rigorous testing
they deserve to determine their long-term positive and/or
negative effects. A recent publication (16) points out
that the color of turmeric is pH dependent. Since
turmeric has been widely used as a coloring agent for
tapa, the use of alkaline buffering agents should be
avoided on the chance that the color of the tapa could be
changed with its application.

Crease Removal. It has been pointed out that a tapa,
however fine, is not completely flat due to the
manufacturing process. In addition, creases from long-
term folding often result in significant deformation
which cannot be entirely removed even with a complete
wetting followed by drying under tension.

The following methods have been suggested by Firnhaber
(University of London Institute of Archaeology Library,
unpublished) for crease removal: (a) "steaming" by
moistening a terrycloth towel, placing the towel on the
crease and ironing at approximately 100 degrees C., (b)
misting the creased area and applying local weights, and
(c) wetting the tapa and applying overall weights. A
further suggestion by Munro (17) is the application of
weights to the edges to create tension and spraying
overall. Steaming and local application of moisture and
weights were considered to be unsuccessful.

Because of the problem of fugitive dyes, alternatives
to the need to apply moisture directly to the tapa have
been considered. Of particular success is a method
devised by Edith Dietze (Smithsonian Institution,

personal communication, 1985) which involved misting of unbuffered archival tissue with distilled or deionized water and the application of the tissue to the back of the tapa with overall weights. A method developed in the Materials Conservation Laboratory of Texas Memorial Museum involved the use of a commercial environmental chamber. The relative humidity was gradually increased in the chamber until the tapa could be manipulated. The edges of the tapa were then placed under weights to create sufficient tension to allow the creases to lessen. While this method prevents any movement of fugitive colors, it is difficult to manage for larger pieces.

Backing and Tear-repair. The methods and materials proposed for lining include the following (a) mulberry paper attached with starch (rice or wheat), methylcellulose, or carboxymethyl-cellulose adhesives (18;19), (b) contemporary tapa attached with starch or cellulose-derived adhesives (20;21), (c) nylon laminating tissue and heat-set polyamide resin (22) and (d) stitched backings.

Backings are usually considered to be a treatment of last resort, used only for tapas which are structurally insecure and would not be adequately supported with patch-type repairs. One problem with a complete backing is that it obscures previous repairs which may be historically significant, particularly if the repair material is tapa applied in a native cultural context. It has also been pointed out that where museum records are incomplete, the source of old repairs may be difficult to attribute if the repairs are of sufficiently similar materials to the tapa (23).

The difficulty in attributing the source of a repair or backing suggests the need for caution in using a contemporary tapa for support. This method was suggested as an alternative to mulberry paper because the new tapa is more easily conformed to the surface irregularities of the tapa being treated than is mulberry paper. For this reason, the contemporary tapa backing should provide greater support than the mulberry paper which would have to be stretched or creased to provide overall, complete contact with the tapa surface. It was further reasoned that the contemporary tapa would look sufficiently different than the tapa being treated to be instantly recognizable as a modern addition. For Fijian tapa, this is probably true, but since modern ethnographic Samoan tapas are frequently backed with pieces of older tapas, a newly manufactured tapa repair might be misleading.

The major disadvantage of applying an overall backing is the loss of flexibility to the tapa cloth. For tapa garments, backings affect drape and three-dimensional form. The suggestion of nylon laminating tissue applied

by heat-set polyamide resin overcomes the disadvantage of
the heavier paper or tapa backings. However, this
introduces a chemically dissimilar material, the ageing
characteristics of which are inferior to either tapa or
mulberry paper, and therefore cannot be recommended.
    Stitching to an auxiliary support is the least
desirable of the methods that have been tried. In one
case observed by Toby Raphael (National Parks Service,
personal communication, 1985), a tapa garment had to be
completely dismantled for the procedure, and stitching
was done through the body of the cloth as well as at the
edges through the original seams. While this treatment,
undertaken at some time in the past, reflects the
application of textile techniques to tapa, it is clearly
a more interventive treatment than would be recommended
today.
    There is a general consensus among the authors cited
here, that tear and crack repair and fills of small voids
can be accomplished with mulberry paper reinforcements.
Larger voids, however, cannot easily be filled by
patching, and are often taken care of by backing. There
are two problems which need to be overcome if a large
void is to be filled by patching: (a) the paper must be
of sufficient weight so that it will not buckle under the
weight of the surrounding tapa, and (b) the fill needs to
be light enough so that it does not strain the weakened
edges of the tapa to which it is being adhered. To a
degree, the edges of a heavier-weight paper can be
chamfered to overcome strain. Alternatively, a lighter-
weight paper can be built up in layers to prevent
buckling and approach the thickness of the tapa. Fills
can be toned by coloring the paper before the repair or
by inpainting after the repair is complete.

Adhesives. The types of adhesives mentioned most
frequently for adhering backings and patches include
starch (both wheat and rice), methylcellulose, and sodium
carboxymethylcellulose. Any of these have good working
properties, sufficient removability, and long-term
durability to make their choice a matter of preference.
However, in an effort to distinguish between the starches
used for original manufacture and the adhesives used in
the repair process, it may be preferable to use a
cellulose ether. This may be especially important in the
repair of delaminations. Here again, there are distinct
disadvantages to using these adhesives with tapas where
fugitive pigments are a problem. Cellulose ethers dry
slowly, and their application adjacent to a fugitive
pigment may cause bleeding.

Filling Voids with Free-Hand Cast Paper Pulp
The use of paper pulp for the filling of voids in paper
has become a standard method of treatment (24;25;26;27).

While the purpose of this technique, when applied to
documents or works of art on paper, is to provide a
uniform, even sheet, the pulp can be purposely left
coarse to imitate the texture of a tapa.
   Equating the techniques developed for use on tapa with
those designed for art of paper represents a rather broad
comparison since leaf casting/manual application of pulp
either does not include the use of adhesives or includes
them in very small amounts (28). In the technique
developed for tapa, the paper is ground with adhesive and
a small amount of water in a blender to make a paste
which is then manipulated into the void with a small
spatula. Test fills have been formed with pulp using
both wheat starch and methylcellulose adhesive. It has
been observed that the wheat starch/pulp mixture can be
used effectively in a nearly dry state, making it ideal
for application to areas adjacent to fugitive colors on
tapa. The pulp/methylcellulose mixture is not easily
worked and tends to break apart when it is used in a
paste dry enough so that it will not affect the movement
of color.
   Robert Futernick has pointed out that experimentation
is necessary to create appropriate pulp for fills for
works of art on paper (29). The same is true for cast
fills for tapa. For example, a soft, dense fill can be
creased using acid-free blotting paper. A thin, fairly
stiff fill was needed for a Samoan tapa (case study of
TMM 1821-18), and this was prepared with Sikishu Kozogame
Mare paper mixed with methylcellulose.

Application of the Vacuum Suction Table

Between October 1986 and the end of September, 1987, 34
tapas from the collections of the Texas Memorial Museum
were treated as part of a project funded in part by the
Institute of Museum Services. Some of the treatment
aspects that had been proposed as part of that project
included: (1) attempting to humidify the tapa by using
the ultrasonic humidifier under the acrylic dome of the
vacuum suction table, using the suction feature to assist
in removing creases; (2) increasing humidification by use
of the ultrasonic humidifier under the acrylic dome,
placing blotters under the tapa to attempt to encourage
the movement of soils and acidic decomposition products
from the tapa to the blotters; (3) attempting to spot
clean soils and stains by application of solvents to the
tapa using the suction feature of the table; (4)
perfecting the application of cast pulp fills for voids
through the use of the vacuum suction table. With the
range of problems presented by this group of tapas, each

of these treatment options was investigated with the
following results:
   1. For small objects that could be accommodated under
the acrylic dome, humidification and crease removal was
significantly faster than by the use of either an
environmental chamber (Hotpak constant
temperature/relative humidity chamber) or an informally
constructed "chamber" consisting of a polyethylene tent
with the tapa suspended over a water source). With the
use of the ultrasonic humidifier, the tapa could reach
the point where it was essentially saturated with
moisture without becoming wet. This was an important
consideration for tapas which were decorated with
pigments that were not stable to wetting. For a small
tapa which fit the space under the dome, the tapa could
be humidified, and at the point where saturation was
nearly complete, the vacuum suction table engaged to pull
the moisture through the tapa while flattening the tapa
to the table surface. By slowly reducing the amount of
humidification until ambient conditions were reached
while under suction, the procedure by which the tapa was
originally processed is imitated. This procedure was
extremely successful except when creases were extremely
sharp. In these instances, however, it was often
impossible to remove those creases completely, even by
wetting the tapa and drying it slowly under weights.
This technique was less successful for pieces which were
larger than the suction table's surface, and therefore
had to be flattened in sections. Two different
techniques were tested for oversized pieces. These
included: (a) extending the acrylic dome by creating a
polyethylene tent over the section of the tapa which was
not on the surface of the suction table, and (b) rolling
the excess area of the tapa over a tube and retaining it
within the dome. Neither method was successful.
Condensation of moisture tended to occur at the join
between the dome and the polyethylene tent, causing drops
of water to form and fall on the surface of the tapa.
While this was solved by containing the tapa within the
dome by rolling the excess over a tube, the tapa would
quickly take on the shape of the curve of the tube,
requiring that the tapa be laid out on a surface after
humidification for further flattening under weights.
This was substantially more time consuming than the
techniques which had been used prior to experimentation
with the suction table, and so was not considered to be
of any advantage.
   2. There had been some hope that it would be possible
to reduce acidity in some of the tapas which could not be
immersed in water by saturating the tapa with moisture
using the ultrasonic humidifier under the acrylic dome,
and moving that moisture through the tapa with the

suction action. This process was described by Marilyn
Weidner in her presentation "Water Treatments and their
Uses Within a Moisture Chamber on the Suction Table"
(30). While some testing of this technique was
undertaken, it was not observed to be successful. There
was limited transfer to the blotters, and it was not
immediately clear whether this was due to the technique
being unsuitable for this application, or that the tapa
was not left for a sufficient time under conditions of
high humidity and suction. Clearly, however, this
technique should have promise for these materials, and it
therefore deserves more rigorous testing accompanied by
some measurement of improvement.
     3. The removal of soils and stains by application of
solvents to the tapa using the suction feature of the
table was not successful. A Bolivian tapa hood with a
painted decoration was chosen for testing of stain
removing capabilities. The hood had been defaced with
green felt-tip marker ink, which proved soluble in a
variety of solvents, including both acetone and water.
It was found that the acetone evaporated too quickly for
the ink to be drawn completely through the tapa onto
blotters below. While the use of water proved more
effective in moving the ink, in neither case was the ink
thoroughly moved into the blotters. Instead, it tended
to remain embedded deep in the fibers, away from the face
of the hood so that it was less visible, but none-the-
less still extant in the tapa fabric.
     4. The most successful application of the vacuum
suction table to the treatment of tapa cloth clearly was
the formation of cast fills for filling voids. As with
the hand-cast fills, pulp was created from hand-made
Japanese papers. The benefit of using the suction table,
however, was that no adhesive was necessary to attach the
pulp to the edges of the voids.
     Trial and error was necessary to achieve the correct
depth and texture for the fill. In addition, it was
necessary to quickly reduce the suction as soon as the
fill was correctly placed to avoid texturing the fill
with the imprint of the suction table screen. The fills
were dried under pressure sufficient only to hold the
fill and surrounding tapa flat as moisture evaporated.
     Difficulties were encountered in the in-painting
process for the fills cast on the suction table. Since
these fills were accomplished without adhesive, they were
even more sensitive to dimensional change by application
of moisture than flat paper or hand-cast mends. Some
experimentation was done in pre-coloring the pulp with
dry pigments. Although this coloring process was
extremely effective, it produced a flat, even color which
was not aesthetically pleasing.

Conclusion

It is clear from this review of techniques suggested for
the treatment of tapa cloth that there is a great deal of
room for further research to determine the effectiveness
of wet treatments, cleaning techniques, the potential for
the application of buffering compounds, and repair
techniques. It is also evident that because there are
substantial differences in the species of plants used,
coloring agents applied and manufacturing techniques
employed, that there must be careful evaluation of tapas
before any conservation treatment can be undertaken.

Case Studies

The following are summaries of treatment reports for
Document 55 (Fiji Museum), 1821-18 (Texas Memorial
Museum), and un-numbered tapa (Department of Textiles &
Clothing, University of Texas at Austin).

Document 55. [Note: this report is similar in concept
to the ideas presented by Natalie Firnhaber in the report
she produced in 1979 (University of London, Institute of
Archaeology Library, unpublished), although the materials
chosen for experimentation are different. As could be
expected from these two early sets of experiments in the
repair and mounting of tapa cloth, the choices of
materials differ from those which might be used at this
point in time.]
    Document 55 is composed of two copies of a single page
of the Samoa Weekly Herald printed on tapa, and dated
March 3, 1900. One page is in three pieces (A, B, C) and
the second page is complete. There are additional copies
of this newspaper in the collections which are in
excellent condition. As a consequence, these pieces have
been given to the conservation department for the purpose
of developing mounting techniques for tapa which can be
applied to masi kesa (painted tapa) in the collections.
    Pieces A, B, and C are folded together and have at some
time in the past been in a fire which burned two sides of
the tapa. When unfolded, the pieces have a generally
oval shape and are approximately 30 cm in width. These
pieces together form a single newspaper page. All three
pages are of a high quality, finely felted tapa which
shows no separation between the layers. The edges of the
tapa are darkened and slightly brittle, and the body
shows some yellowing. The tapas were tested to determine
their pH with a flat electrode. Measurements on the
three pieces ranged from 4.5 to 5.2.
    Piece D is a full sheet of newspaper (tapa) measuring
59 x 82cm. The edges are somewhat irregular, and the
tapa is not of good quality, being poorly felted and thin

in some areas. There is some stretching along the thin
areas which had blurred the printing. The document
ranges in color from yellow to brown, and pH measurements
with a flat electrode ranged from 4.0 to 4.5.
Treatment proposal
    1. The problem posed is as follows: a) reduce the
acidity in the tapas to prolong their potential for
preservation, b) choose an appropriate backing material
to act as a support and enable the documents to be
handled without endangering their fragile edges, and c)
choose an appropriate adhesive to attach the tapa to the
backing material.
    1.A. Deacidification: test inks for stability to
water and methanol. If the inks are stable to water, the
tapas will be immersed in deionized water and deacidified
using the soda siphon method (Clapp, A., 1978.
Curatorial Care of Works of Art on Paper, Intermuseum
Conservation Association, Oberlin, Ohio, p.73). If the
inks are not stable in water but are stable to methanol,
the tapas will be treated with WEI T'O Solution #3. If
the inks are unstable in both water and methanol, the
tapas will be encapsulated in Mylar polyester film and no
further treatment will be attempted.
    1.B. Choice of Backing Material: The following
materials are available in the laboratory to use as
backings for the tapas - contemporary tapa, commercial
(machine made) Japanese tissue of very light weight, and
Cerex (nylon 6.6 type) web fabric. The Japanese tissue
is too thin to provide adequate support for the tapas and
has been eliminated from the choices. Tapas A-C will be
mounted on the contemporary tapa and Tapa D will be
mounted on the Cerex.
    1.C. Adhesives: The adhesives available in the lab
for mounting the tapas include the following types - PVA
emulsion (Vinamul 5816, Jade 403, PVA from Process
Materials Corp.), PVB resin (Hoechst), Polyamide web
adhesive (Process Materials Corp.), Methyl cellulose
paste (Process Materials Corporation), and wheat starch
paste (Talas).
    Test squares of contemporary tapa and Cerex were cut
and adhered together to test the workability of each of
the adhesives with the following results: (1) the PVA
emulsions were insufficiently reversible after drying,
(2) the PVB, used as a heat-set adhesive tended to come
through to the front of the tapa when the two tapas were
adhered; lesser amounts were used but a good bond could
not be produced, (3) polyamide web was successful both
for adhering tapa to tapa and for adhering the Cerex to
the tapa and remained readily soluble in methanol, (4)
methylcellulose and wheat starch were both successful and
easily reversible.
    Because of the historic reliability of starch

adhesives, wheat starch was chosen for the mounting of
Document 55 A-C to the contemporary tapa. Polyamide web
was the only adhesive available in the lab which would
adhere the tapa of Document 55-D to the Cerex.
    2. Treatment of Document 55 A-C.
    2.A. Deacidification: Inks were tested for water
solubility and were found to be stable even with
prolonged contact with moisture. The tapas were
supported on nylon screens and immersed in three changes
of deionized water. Flat electrode readings after each
water change showed a decrease in acidity to 5.5 for 55 B
and 6.0 for 55 A and C. The tapas were blotted with
acid-free blotter paper to remove excess moisture and
then treated with magnesium bicarbonate according to the
directions in Curatorial Care of Works of Art on Paper.
The final pH was measured at 6.5 The tapa was dried on
changes of blotter paper under glass weights.
    2.B. Mounting: A 4-layer section of contemporary
white tapa was cut to the size of the original newspaper
page (based on the measurements of 55 D) and deacidified
in the same manner as the artifact tapa. The tapa pieces
were placed face down on release paper and starch
adhesive applied to the back. The contemporary tapa was
centered over the artifact tapas, and the assemblage
placed under glass weights to dry.
    3. Treatment of Document 55 D.
    3.A. Deacidification: The same procedure was used for
Document 55 D as had been used for Document 55 A-C.
    3.B. Mounting: Heat set tissue was cut to match the
size and shape of the artifact tapa, and spot-tacked into
place with a tacking iron. The document with its
adhesive backing was centered on the Cerex sheet, and
placed between layers of silicone release paper. The
assemblage was placed in a dry-mount press at 150 degrees
F. for 20 seconds. The assemblage was removed from the
press immediately, the silicone paper discarded, the
mounted tapa placed under weights for 20 minutes.
    Evaluation: The tapa-mounted artifact produced the
most satisfactory results. The Cerex backing proved to
be too insubstantial and had to be mounted to a piece of
acid-free mat board in order to be handled. (Completed
2/80.)

1821-18 - Samoan Tapa (Texas Memorial Museum)
    Dimensions: 79 x 148 cm
    Description: Beaten bark cloth of varying thickness.
The overall background design has been produced from a
printing board and features zig-zag lines and short
diagonal hatched lines in a light reddish-brown.
Overpainting of some of the zig-zag lines in done in dark
brown, and typical Samoan circles and half-moon shapes
are scattered across the surface.

Condition: The tapa is in poor condition; the fabric being dry and extremely brittle. There are numerous tears and holes, some delaminations and some loss. Losses are severe along the dark brown border which is significantly more brittle than other areas of the tapa. There is a fairly substantial area of loss at the center of one of the shorter ends. There are several detached fragments and a small pile of fibers in the storage container. The tapa is creased from being stored folded.

Proposal of Treatment:

1. Test colors for fastness to water. If colorfast, humidify under tension to relax creases.

2. The tapa is too fragile, and the edges too irregular to be handled safely without a complete backing. Because of the light, rather stiff character of the tapa, Sikishu Kozogami Mare paper is proposed for this purpose.

3. Fill voids with pulp made of the same paper as the backing, inpainting with acrylic polymer paints to blend.

Treatment employed:

1. The fiber residue in the storage bag was tested as pH 4.4 with an Orion Research Ionanalyzer Model 407A using cold extraction techniques. No color from the dyes was evident in the water, but the water became quite yellow.

2. A small fragment which could not be associated with a specific location on the tapa was immersed for 10 minutes in deionized water. A significant amount of yellow material was visible in the water, but the dark brown and reddish brown colors appeared stable. The fragment was blotted and dried under fresh blotters and weights. After 15 minutes some movement of the dark brown color was observed on the blotters. When dry, there was a marked improvement in the flexibility of the tapa, and a decision was made to float-wash the larger piece after it had been humidified to flatten the creases and provide overall, even contact between the tapa and the water.

3. The tapa was flattened by placing in a humidification chamber constructed of polyethylene over a work table. Moisture was introduced by misting within the chamber. The tapa was placed face down on blotters with weights at the edges to keep the piece under slight tension. Flattening took less than 2 hours.

4. The flattened tapa was float-washed in deionized water for 5 minutes per change of water using 4 changes.

5. The tapa was transferred to blotters (face up and uncovered), and the blotters changed frequently over the course of an hour. No movement of the dark brown color was observed. The tapa was placed on silicone release paper face up and weights placed at the edges to provide tension, and drying was completed (silicone release paper

was substituted for blotters to prevent the wicking of
any pigments into the blotters through the back of the
tapa).
    6. Sikishu Kozogami Mare paper was pasted out with
methylcellulose and applied to the back of the tapa. The
backed tapa was placed face up on blotter paper and the
edges weighted to provide tension until the backing was
dry. A small amount of moisture from the adhesive was
evident on the blotters, but did not come through to the
face of the tapa.
    7. Paper pulp was made from the Sikishu Kozogami Mare
paper by soaking in deionized water, squeezing excess
moisture out with blotters, and then processing for a few
seconds in a blender with methylcellulose paste to
produce a coarse pulp for filling the voids to the level
of the face of the tapa. The pulp was placed in the
voids with a spatula, the area surrounding the fill area
weighted to prevent the edges from curling, and the fill
allowed to dry. The fills were then inpainted with
acrylic polymer pigments (Liquitex) to blend.
    Evaluation: The backing caused a minimal increase in
stiffness. The pulp fills were effective in filling the
voids, and from an aesthetic viewpoint were more
appropriate than the flat fills produced by the backing
alone. (completed 1/86)

No Number Tapa (Department of Textiles and Clothing,
University of Texas at Austin)
    Dimensions: unknown at start of treatment. The tapa
is folded, apparently in thirds and then in quarters at a
size of approximately 17" x 23".
    Description: Beaten bark cloth of varying thickness.
The overall design in reddish brown and dark brown
geometric designs appear to be rubbed designs produced by
a printing board. Some areas appear to have been over-
painted with a darker brown.
    Condition: The tapa is in extremely fragile and
brittle condition. It cannot be unfolded without
potentially causing further damage. There are cracks in
the fold lines evident, as well as ragged edges. Cold
extraction yielded a pH of 5.7
    Proposed treatment:
    (1) humidify to allow fibers to relax sufficiently so
that the tapa can be unfolded. Test for pH and fastness
of colors to immersion in water.
    (2) place weights along folds of humidified tapa to
ease out creases.
    (3) temporarily align breaks and fasten in place with
temporary mends of Japanese paper (Tengujo) with
methylcellulose adhesive.
    (4) fill voids with Tengujo pulp applied on the vacuum
suction table.
    (5) inpaint mends and fills with acrylic pigments.

Treatment employed:
1. A disassociated fragment was tested as pH 5.7 with
an Orion Research Ionanalyzer Model 407A using cold
extraction techniques.
2. Coloring agents were tested for fastness to
immersion in water, and all were found to be fugitive.
No wet treatment was considered.
3. The tapa was placed in a polyethylene tent and the
humidity raised using an ultrasonic humidifier. As the
fibers relaxed, the tapa was unfolded and the creases
eased out under weights.
4. Breaks and tears were aligned and mended both from
the front and back with wet-torn Japanese paper (Tengujo)
using methylcellulose adhesive. The face mends were
removed after repairs and fills were complete.
5. A thin, highly processed pulp was made by soaking
Tengujo paper in deionized water and then processing in a
blender. The tapa was placed on the suction table with
all but the void areas masked out with the vinyl covering
blankets. With the suction feature turned on, pulp was
fed into the voids with a large bore syringe. As soon as
the fill was complete, the suction was reduced to a level
which would do no more than hold the area flat until it
dried.
6. After all of the voids were filled, inpainting of
the mends and fills was accomplished with Liquitex
acrylic paints.
Evaluation: In spite of the difficulties of working
pigments into a fill created on the suction table without
an adhesive binder, this method produces the best results
of all methods previously tested. The repair process is
rapid, and fills can be easily identified from the
reverse of the object. In addition, the fills are
removable by slight wetting if that should become
necessary.

Acknowledgments

Meredith Montague, Assistant Conservator, Materials
Conservation Laboratory, Texas Memorial Museum, carried
out the replication of the experiments done with the
vacuum suction table, and completed the treatment of the
unnumbered tapa from the University's Department of
Textiles and Clothing.

Literature Cited

1. Green, S.W. Preprints of Papers Presented at the
    Fourteenth Annual Meeting, A.I.C., 1986, pp 17-31.
2. Turchan, C. Abstracts: The Care and Preservation of
    Ethnological Materials: Symposium '86, Ottawa, 1986.

3. Firnhaber, N.; Erhardt, D. Abstracts: The Care and
   Preservation of Ethnological Materials: Symposium
   '86, Ottawa, 1986.
4. Erhardt, D.; Firnhaber, N. Recent Advances in the
   Conservation and Analysis of Artifacts, University of
   London, Institute of Archaeology Jubilee Conservation
   Conference, 1987, pp 223-237.
5. Ling,S.; Ling, M. Bark Cloth, Impressed Pottery, and
   the Inventions of Paper and Printing. Ins. Ethnology
   Academia Sinca: Nanking, Taipei, Taiwan, 1963.
6. Kooijman, S. Tapa in Polynesia. Bishop Museum Press:
   Honolulu, Hawaii, 1972.
7. Kooijman, S. Tapa in Polynesia. Bishop Museum
   Press: Honolulu, Hawaii, 1972, p 347.
8. Kooijman, S. Tapa on Moce Island, Fiji. Rijksmuseum
   voor Volkenkunde: Leiden, 1977.
9. Leonard, A.; Terrell, J. Patterns of Paradise.
   Field Museum of Natural History: Chicago, 1980.
10. Kooijman, S. Tapa on Moce Island, Fiji. Rijksmuseum
    voor Volkenkunde: Leiden, 1977, p 216.
11. Wolf, S. Texas Memorial Museum Conservation Notes,
    1983, 5.
12. Leonard, A.; Terrell, J. Patterns of Paradise, Field
    Museum of Natural History: Chicago, 1980, p 17.
13. Leonard, A.; Terrell, J. Patterns of Paradise, Field
    Museum of Natural History: Chicago, 1980, p 18.
14. Bakken, A.; Aarmo, K. ICOM Committee for
    Conservation 6th Triennial Meeting Preprints,
    1981, p 81/3/2-4.
15. Bakken, A.; Aarmo, K. ICOM Committee for
    Conservation 6th Triennial Meeting Preprints,
    1981, p 81/3/4.
16. Lee, D.; Bacon, L.; Daniels, V. Studies in
    Conservation, 1985, 30, p 184-188.
17. Munro, S. ICOM Committee for Conservation 6th
    Triennial Meeting Preprints, 1981, p 81/3/3.
18. Wolf, S.; Fullman, G. ICCM Bulletin, 1980, 6.
19. Wolf, S. Texas Memorial Museum Conservation Notes,
    1983, 5, p 2.
20. Wolf, S.; Fullman, G. ICCM Bulletin, 1980, 6.
21. Wolf, S. Texas Memorial Museum Conservation Notes,
    1983, 5.
22. Munro, S. ICOM Committee for Conservation 6th
    Triennial Meeting Preprints, 1981, p 81/3/3.
23. Munro, S. ICOM Committee for Conservation 6th
    Triennial Meeting Preprints, 1981, p 81/3/3-4.
24. Keyes, K.; Farnsworth, D. Preprints of the 4th
    Annual Meeting, A.I.C., 1976, pp 76-80.
25. Futernick, R. Journal of the A.I.C., 1983, 22,
    pp 82-91.
26. Perkinson, R.; Futernick, R. In: Preservation
    of Paper and Textiles of Historic and Artistic

Value, Williams, J., Ed. ACS Series 164; American
Chemical Society, Washington, DC, 1977, pp 102-111.
27. Petherbridge, G. Conservation of Library and Archive
Materials and the Graphic Arts, Cambridge, 1980
Preprints, 1980, pp 189-209.
28. Keyes, K; Farnsworth, D. Preprints of the 4th
Annual Meeting, A.I.C., 1976, pp 76-80.
29. Futernick, R. Journal of the A.I.C., 1983, 22,
pp 82-91.
30. Weidner, M. Preprints of Papers Presented at
the 13th Annual Meeting, A.I.C., 1985.

RECEIVED April 24, 1989

CHARACTERIZATION AND PRESERVATION OF TEXTILES

Chapter 13

Identification of Red Madder and Insect Dyes by Thin-Layer Chromatography

Helmut Schweppe[1]

BASF Aktiengesellschaft, D–6700 Ludwigshafen,
Federal Republic of Germany

Red natural anthraquinone dyes on ancient textile materials
can be readily identified by thin-layer chromatography (TLC)
if they belong to the class of madder dyes. The method also
shows which type of dye plant from the family Rubiaceae has
been used for dyeing (Rubia tinctorum L., R. peregrina L.,
R. cordifolia L., R. akane, various Galium spp., Relbunium
spp., Morinda spp., Oldenlandia spp., Coprosma spp., or Ventilago spp. (Rhamnaceae)). Any changes in the composition
of the dyes during its extraction from the dyed material, e.g.
the transition from pseudopurpurin to purpurin, can be preven-
ted by suitable preparation of the sample before TLC.
      The red insect dyes from Dactylopius coccus COSTA (Ame-
rican cochineal), Kermococcus vermilio PLANCHON (kermes), and
Kerria lacca KERR (lac dye) can also be readily distinguished
by thin-layer chromatographic comparison.
      Porphyrophora polonica L. (Polish cochineal) contains
small amounts of the kermes dyes kermesic acid and flavokermesic acid besides the cochineal dye carminic acid. These secondary components cannot be identified unless they have previously been concentrated.
      Porphyrophora hameli BRANDT (Armenian cochineal) contains
nearly exclusively carminic acid. It has been reported that
high-performance liquid chromatography (HPLC) has also identified a trace of kermesic acid, but TLC has not provided any
clear proof of the presence of this secondary component in Armenian cochineal, even after previous concentration.

Besides indigo and Phoenician or Tyrian purple, the red madder and
insect dyes were of particular importance for the dyeing of textile
materials in earlier centuries. These dyes were used to produce the highly
prized red and violet dyeings with outstanding fastness to light and

[1]Current address: Paul-Klee-Strasse 11, D–6710 Frankenthal, Federal Republic of Germany

0097–6156/89/0410–0188$9.00/0
© 1989 American Chemical Society

washing. Owing to their good lightfastness, these dyed shades have
remained almost unchanged through the centuries. There are several
very old textile objects to prove this, for example the Pazyryk carpet, the oldest knotted carpet in the world (1), many pre-Columbian
Peruvian (2,3,3a) and Coptic textiles (4) that are still in good
condition, and the Sicilian coronation robe of the Hohenstaufen Emperors (2).
     Many dyeings produced with other natural dyes have far inferior
lightfastness, and many of them have faded or changed in shade.
A good example is the famous "Polish carpets" (5), whose silk pile
has been dyed partly with safflower carmine (C.I. Natural Red 26) *,
which has poor lightfastness. When we remove the dyes from such dyeings with poor lightfastness, we obtain extracts that contain photochemical degradation products of the dyes as well as the dyes themselves. The degradation products sometimes produce additional spots
on thin-layer chromatograms and additional peaks on the chromatograms in high-performance liquid chromatography. This may cause
trouble in evaluating the chromatograms.
     When the historic textile material to be investigated contains
a red that has hardly faded, we should start by testing this red dyeing, because this is the simplest and quickest way to obtain a result. In the first place, we need a smaller sample for the identification of red anthraquinone dyes than for identifying yellow flavone
dyes, and in addition, it is then usually possible to identify the
dyer's plant or dye insects that have been used for the red shade.
In contrast, dyer's plants belonging to the hydroxyflavone class often contain the same main components, so that it is not possible to
identify the individual plant that has been used. For instance, the
hydroxyflavones quercetin (C.I. 75670) and luteolin (C.I. 75590) are
contained in many dyer's plants as the main component or the sole
dye.
     Significant progress has recently been made in the identifica-
tion of red madder and insect dyes on historic textile materials
(2,7,8,9). As the most of these natural dyes are mixtures, chromatographic methods such as TLC and HPLC are the preferred techniques.
It may have been the exhaustive study by Donkin (10) on insect dyes
that prompted some experts to return to the problem of distinguishing
between insect dyes. In the light of the results obtained by Donkin
and the data presented by Taylor (8) on the revised zoological names
for the dye insects, we must now distinguish between the following
five dye insects that were used in the past for dyeing textile materials:

Dactylopius coccus COSTA, which produces American cochineal;
Kermococcus vermilio PLANCHON, which produces kermes;
Porphyrophora polonica L., which produces Polish cochineal;
Porphyrophora hameli BRANDT, which produces Armenian cochineal;
Kerria lacca KERR, which produces lac dye (8,11).

---

* In the "Colour Index" (C.I.), a multiple-volume, English reference
work (6), the names, commercial denominations, constitutions, and
dyeing properties of synthetic and natural dyes are listed. Each dye
has a generic name, and, if the constitution is known, a constitution
number.

Of these natural dyes, cochineal and lac dye[*] are the only products
that are available in the market. It is very difficult to obtain
samples of the other three dye insects.
The separation of insect dyes and madder dyes by HPLC is discussed exhaustively by Wouters (7). Various other laboratories (8,
12-14) also use this method for separating natural dyes. There are
far more publications, however, on the use of TLC for separating
natural dyes, including those belonging to the class of hydroxyanthraquinones (4, 8, 9, 12-31).
A comparison of the two chromatographic methods shows that they
have the following advantages and disadvantages:
HPLC has outstanding separating efficiency; the retention times (comparable with the R_f values in TLC) are well reproducible, and only
very small amounts of material are required for separation; the UV
detector is highly sensitive and identifies secondary components even
in very low concentrations. HPLC also provides an approximate survey
of the quantitative composition of the sample tested. A significant
disadvantage of HPLC, as compared with TLC, is the method of identification by means of the "blind" UV detector, which records the components of a mixture on the chromatogram in the form of "peaks" differing in height merely according to the amount of material present.
In contrast, TLC of dyes offers various methods of identifying
the components of a mixture one after the other. First of all, we can
compare the inherent colors of the dye spots, and then the fluorescent colors in UV light, and finally, we can produce the uranyl, aluminum, zirconium, magnesium or calcium lakes by dipping the chromatogram in inorganic salt solutions of these substances. The uranyl
lakes of hydroxyanthraquinones show the largest differences in their
shades (2). In separating efficiency, reproducibility of the retention times or R_f values, and the identification limit for secondary
components, TLC is inferior to HPLC.
In future, HPLC will be used on an increasing scale for separating complicated mixtures of natural dyes. It is possible that a
combination of HPLC with FTIR spectroscopy will help to overcome the
"blindness" of the UV detector.
TLC will continue to maintain its position for the separation
of hydroxyanthraquinones, particularly in the smaller laboratories
that cannot afford the expensive equipment for HPLC.

DYES IN DYER'S PLANTS OF THE TYPE OF MADDER

In recent decades, numerous dyer's plants of the type of madder
(Rubia tinctorum L.) have again been investigated for their constituents. With the aid of modern methods of instrumental analysis, no
less than 23 different components belonging to the class of hydroxyanthraquinones have been identified in madder roots (32-34), for instance, besides five hydroxyanthraquinone glycosides, of which rubianin (Constitution VIII in table I) (35) is of particular interest,

---

[*]Supplier: Mann (Natural Dyes), Im Dorngarten 6, D-6719 Lautersheim,
Federal Republic of Germany; Tel. 06351/6869

because it is a hydrolysis-stable C-glycoside, a constituent that has hitherto only been found in the red insect dye carminic acid (constitution XXVII in table 4). The constitutions of the madder dyes are listed in table I.

Table I. Constitutions of the madder dyes

No.	Name	C.I.Constitution Number (6)	Constitution
I	Alizarin	75330	1,2-Dihydroxyanthraquinone
II	Purpuroxanthin	75340	1,3-Dihydroxyanthraquinone
III	Rubiadin	75350	1,3-Dihydroxy-2-methylanthraquinone
IV	Morindanigrin	75360	1,3-Dihydroxy-6-methylanthraquinone
V	Lucidin	---	1,3-Dihydroxy-2-hydroxymethylanthraquinone
VI	Damnacanthal	---	1,3-Dihydroxyanthraquinone-aldehyde
VII	Munjistin	75370	1,3-Dihydroxyanthraquinone-2-carboxylic acid
VIII	Rubianin	---	1,3-Dihydroxy-2-C-glycosyl-anthraquinone
IX	Quinizarin	58050	1,4-Dihydroxyanthraquinone
X	Christofin	---	1,4-Dihydroxy-2-ethylhydroxymethylanthraquinone
XI	-----	---	1,4-Dihydroxy-2-hydroxymethylanthraquinone
XII	Quinizarin-2-carboxylic acid	---	1,4-Dihydroxyanthraquinone-2-carboxylic acid
XIII	-----	---	1,4-Dihydroxy-6-methylanthraquinone
XIV	Soranjidiol	75390	1,6-Dihydroxy-2-methylanthraquinone

Continued on next page

Table I. Continued

No.	Name	C.I.Constitution Number (6)	Constitution
XV	Physcion	---	1,8-Dihydroxy-6-methoxy-3-methylanthraquinone
XVI	Physcionanthranol A	---	1,8-Dihydroxy-6-methoxy-3-methylanthranol-10
XVII	Physcionanthranol B	---	1,8-Dihydroxy-6-methoxy-3-methylanthranol-9
XVIII	Anthragallol	58200	1,2,3-Trihydroxyanthraquinone
XIX	Anthragallol-2-methyl ether	---	1,3-Dihydroxy-2-methoxyanthraquinone
XX	Anthragallol-3-methyl ether	---	1,2-Dihydroxy-3-methoxyanthraquinone
XXI	Purpurin	75410	1,2,4-Trihydroxyanthraquinone
XXII	Pseudopurpurin	75420	1,2,4-Trihydroxyanthraquinone-3-carboxylic acid
XXIII	Morindon	75430	1,5,6-Trihydroxy-2-methylanthraquinone
XXIV	Emodin	75440	1,6,8-Trihydrox-3-methylanthraquinone
XXV	Copareolatin-dimethyl ether	---	6,8-Dihydroxy-4,7-dimethoxy-3-methylanthraquinone or 4,6-Dihydroxy-7,8-dimethoxy-3-methylanthraquinone

Rubia tinctorum L. (Rubia tinctoria SALISB.) (C.I. Natural Red 8)

Indigenous in South and Southeast Europe, in the Mediterranean area, in Asia Minor and in the Caucasus, and from there to China and Japan and Malaysia, and in the west part of North America, Mexico, and South America (34).

In the past, madder was cultivated in huge amounts for the production of dyes from its roots, but today it is only found growing wild in Asia Minor. Madder has always been one of the most important of all dyer's plants, and it was used as early as 2000 B.C. for dyeing textiles (36), or even earlier (37).

Constituents in Radix Rubiae (madder roots): 2-3,5% of di- and trihydroxyanthraquinone glycosides, and as the main constituent ruberythric acid (alizarin-2-ß-primveroside). Also galiosin (pseudopurpurin primveroside), rubiadin-3-ß-primveroside, rubiadin-3-ß-glucoside, and lucidin-3-ß-primveroside (38), and rubianin (VIII), a C-glycoside (35). Madder root contains as free hydroxyanthraquinone dyes pseudopurpurin (XXII), rubiadin (III), alizarin (I), and munjistin (VII), and also small amounts of christofin (X) (32, 34, 39).

The dried root was also found to contain purpuroxanthin (II) formed by separation of CO_2 from munjistin (VII). Pseudopurpurin (XXIII) is probably decarboxylated with formation of purpurin (XXI) when the madder root is dried. The dried madder root also contains the following hydroxyanthraquinone dyes in small amounts: nordamnacanthal (VI), quinizarin (IX), 1,4-dihydroxy-2-hydroxymethylanthraquinone (XI), quinizarin-2-carboxylic acid (XII), anthragallol (XVIII), and anthragallol-3-methyl ether (XX) (32, 34, 39).

The other hydroxyanthraquinone constituents identified in madder root have only one free phenolic OH group and are not, therefore, suitable mordant dyes for textile materials. They will be disregarded in the following. This leaves us, therefore, with only 15 of the 23 constituents of madder root belonging to the class of hydroxyanthraquinones (see table II) that can be used as mordant dyes (in the free form or (partly) as glycosides) with varying dyeing properties. As not all of them have good affinity for mordanted textile fibers, and some of them are contained only in small amounts in madder root, we only have to recognize a few of these mordant dyes on the chromatogram in TLC comparisons to identify a madder dyeing. These madder dyes are alizarin (I) and purpurin (XXI) and also pseudopurpurin (XXII) in cases where the latter has not been completely converted by decarboxylation into purpurin.

Rubia peregrina L (R. anglia HUDS., R. lucida L.) (C.I.Natural Red 8)

Wild madder or Levantine madder. Indigenous in the Mediterranean area and in the Orient.

Constituents: The roots of Wild madder contain galiosin, pseudopurpurin (XXII), purpurin (formed in some cases from pseudopurpurin by decarboxylation), and a small amount of alizarin (I) (See table II).

Rubia cordifolia L. (R. cordata THUNB., R. munjista ROXB.) (C.I. Natural Red 16)

East Indian madder. Munjeet. Indian madder.

Indigenous in India, Nepal, China, Japan, and tropical Africa.
In the past, the red root was traded as "East Indian madder root".
Constituents (See also table II): The roots contain alizarin (I),
pseudopurpurin (XXII), purpurin (XXI), purpuroxanthin (II) and munjistin (VII) (34). Nordamnacanthal (VI), physcion (XV), and 1,4-dihy-
droxy-6-methylanthraquinone (XIII) have also been identified (40).
In contrast to Rubia tinctorum, the roots of this plant contain no
lucidin (V).
    Pfister (41) assumes that Rubia cordifolia was used by antique
Indian dyers. Another red madder dye that later was used in India is
the Chay root of Oldenlandia umbellata L. (42).

Rubia akane.(NAKAI) Japanese madder.

    Constituents (See also tableII): The roots contain a glycoside
of pseudopurpurin (XXII) (43-45) that is relatively stable to hydrolysis with dilute mineral acids (46).
    Rubia akane has been identified on fragments of silk fabrics
from coffins with mummies from the Fujiwara era (11th century A.D.)
(47).

Relbunium hypocarpium (L.) HEMSL.. Relbún. Ruivina. Chamri (local name in Chile).

Indigenous in Central and South America.
    Constituents (See also table II): Relbun roots contain galiosin
as the pricipal dye, but pseudopurpurin (XXII) and purpurin (which
may have been formed from pseudopurpurin by decarboxylation) may also
be present (48). Alizarin has not been found in relbun roots (49).

Relbunium ciliatum (L.) HEMSL.

A sample of the roots of this species has also been tested. The dye
composition was found to be the same as that in the roots of R. hypocarpium.

On numerous Peruvian textiles from the Paracas era, dyes from relbun
roots have been identified (2, 3, 48).The Araukans, an Indian tribe in
southern Chile, have been using roots of Relbunium hypocarpium even
in this century for dyeing red shades on wool (50).

Galium verum L. (Lady's bedstraw. Yellow galium); Galium mollugo L.
(G.album. Hedge bedstraw) (C.I. Natural Red 14)

Both galium spp. are indigenous in Europe, and in the area from the
Caucasus to the eastern part of India.
    Constituents (See also table II):The dyes contained in the
roots of these two plants are pseudopurpurin (XXII), its primveroside
galiosin, purpurin (XXI) (formed in some cases from pseudopurpurin by
decarboxylation), rubiadin (III), its 3-ß-primveroside, lucidin (V)
(partly as glycoside), purpuroxanthin (II), and alizarin (I) (partly
as glycoside) (51).
    Galium roots were used in Scotland for dyeing bordeaux shades
on alum mordanted wool (52, 53).They have also been identified on a

fabric found by excavation in the Viking harbor Haithabu (north Germany) (54).

Oldenlandia umbellata L. (O. hispida DC.; Hedyotis indica R.et SCH.)
(C.I. Natural Red 6)

Indigenous in India, Burma , Abyssinia, Ceylon, and Java.
The root was traded under the names Chay root and "Indian Madder"* as a red natural dye.
Constituents (See also table II): The root of Oldenlandia umbellata L. contains alizarin (I) (mostly in the form of its glycoside ruberythric acid) as the sole dye —no pseudopurpurin (XXII) and no purpurin (XXI). Some monohydroxyanthraquinones, for example, hystazarin monomethyl ether (2-hydroxy-3-methoxyanthraquinone) are also present, but they do not go onto the textile material in dyeing, but remain in the dyeing liquor. Dyeings with Chay root have the same brilliant shade as those with synthetic alizarin. They cannot be distinguished from each other by TLC. The roots of O. umbellata were very suitable for dyeing cotton by the Turkey red oil process. In this process, dyes such as purpurin and pseudopurpurin cause trouble, because they impair the good wash fastness of these dyeings.
It is possible that the first Turkey red oil dyeings were produced in India with the Chay root (55).

Morinda citrifolia L. (C.I.Natural Red 18)

Indigenous in India, Indochina, Malaya, Thailand, the Polynesian islands, and the Philippines, and also cultivated in parts of India.
The root or the bark of the root used to be traded as natural dye under the names Morinda root, Suranji, and Al (3).
Constituents (See also table III): Approx. 0,4% of hydroxyanthraquinones: the principal components are Morindon (XXIII) and Morindin (morindon-5-rutinoside). Other constituents are soranjidiol (XIV), morindadiol (1,5-dihydroxy-2(or 3)-methylanthraquinone), rubiadin (III), nordamnacanthal (VI), alizarin (I), a trihydroxymethylanthraquinone monomethyl ether (56) and probably a small amount of emodin (XXIV) (2).

Morinda umbellata L. (C.I. Natural Red 19)

Indigenous in the east part of India, in Ceylon and Java.
The root bark was traded under the names Mang-kouda or Mang-kuda as a natural dye, and it was used in Java for fast red batik prints.
Constituents (See also table III): The root or its bark contains as its principal dyes morindon (XXIII) and morindin (morindon-5-rutinoside). Other constituents are soranjidiol (XIV), morindadiol, rubiadin (III) (partly as glycoside), purpuroxanthin (II), lucidin (5), munjistin (VII), and small amounts of alizarin (I), Morindanigrin (IV) and probably Emodin (XXIV) (2,6,56,57).

---

*As the root of Rubia cordifolia L. was sometimes marketed under the name "Indian madder", it is generally advisable to use the name "Chay root" for the root of Oldenlandia umbellata L. to avoid confusion.

Table II. Hydroxyanthraquinone dyes in Rubia-, Relbunium-, Galium-, and Oldenlandia-species

	A	B	C	D	E	F	G	H
Alizarin (I)*	+	+	+	-	-	+	+	+
Purpuroxanthin (II)	+	-	+	-	-	+	+	-
Rubiadin (III)*	+	-	-	-	-	+	+	-
Lucidin (V)*	+	-	-	-	-	+	+	+
Nordamnacanthal (VI)	+	-	-	-	-	+	+	-
Munjistin (VII)	+	-	+	-	-	-	-	-
Rubianin (VIII)	+	-	-	-	-	-	-	-
Quinizarin (IX)	+	-	-	-	-	-	-	-
Christofin (X)	+	-	-	-	-	-	-	-
1,4-Dihydroxy-2-hydroxy-methylanthraquinone (XI)	+	-	-	-	-	-	-	-
Quinizarin-2-carboxylic acid (XII)	+	-	-	-	-	-	-	-
1,4-Dihydroxy-6-methyl-anthraquinone (XIII)	-	-	+	-	-	-	-	-
Physcion (XV)	-	-	+	-	-	-	-	-
Anthragallol (XVIII)	+	-	-	-	-	-	-	-
Anthragallol-3-methyl ether (XX)	+	-	-	-	-	-	-	-
Purpurin (XXI)	+	+	+	+	+	+	+	-
Pseudopurpurin (XXII)*	+	+	+	+	+	+	+	-

* Partly or entirely as glycoside
+ Dye detected
- Dye not detected

A: Rubia tinctorum L.; B: Rubia peregrina L.; C: Rubia cordifolia L.; D: Rubia akane; E: Relbunium hypocarpium HEMSL.; F: Galium verum L.; G: Galium Mollugo L.; H: Oldenlandia umbellata L.

Coprosma species (Rubiaceae)

The plants of the coprosma species are indigenous in Australia, New Zealand, and on many South Sea islands.

The stem or root bark of various coprosma shrubs contains a number of anthraquinone dyes. It was used by the Maoris, the natives of New Zealand, for dyeing flax in orange shades.
        Coprosma lucida J.R. & G. FORST was called by the natives "orange leaf". They called C. grandiflora HOOK. and C. areolata CHEESEM "karamu", and C. linariifolia HOOK. "yellow wood" (58).
        Constituents (See also table III): The stem and root barks differ considerably in their constituents that are suitable for dye-

ing. These hydroxyanthraquinone dyes of the four coprosma species
mentioned above are listed in table III (58, 59).
    As comparison material, only the bark of coprosma lucida was
available. The dyeing produced with this bark on alum-mordanted wool
has an orange shade.

Table III. Hydroxyanthraquinone dyes in Morinda-, Coprosma-
species, and in Ventilago mad(e)raspatana GAERTN.

	A	B	C	D	E	F	G
Alizarin (I)*	+	+	-	-	-	-	-
Purpuroxanthin (II)	-	+	-	-	-	-	-
Rubiadin (III)*	+	+	+	+	-	-	-
Morindanigrin (IV)	-	+	-	-	-	-	-
Lucidin (V)*	-	+	+	-	-	-	-
Nordamnacanthal (VI)	+	-	-	-	-	-	-
Munjistin (VII)	-	+	-	-	-	-	-
Morindadiol	+	+	-	-	-	-	-
(= 1,5-Dihydroxy-2-(oder 3)-methylanthraquinone)							
Soranjidiol (XIV)	+	+	+	+	-	-	-
Physcion (XV)	-	-	-	-	-	-	+
Physcionanthranol A (XVI)	-	-	-	-	-	-	+
Physcionanthranol B (XVII)	-	-	-	-	-	-	+
Anthragallol (XVIII)	-	-	+	-	-	-	-
Anthragallol-2-methyl ether (XIX)	-	-	+	-	-	+	-
Morindon (XXIII)*	+	+	-	+	-	-	-
Emodin (XXIV)	+	+	-	-	-	-	+
Copareolatin	-	-	-	+	-	-	-
(= 4,6,7,8-Tetrahydroxy-3-methylanthraquinone)							
Copareolatin dimethyl ether (XXV)	-	-	+	-	-	-	-

*Partly or entirely as glycoside
+ Dye detected
- Dye not detected

A: Morinda citrifolia L.; B: Morinda umbellata L.; C: Coprosma
lucida J.R. & G.FORST.; D: C. grandiflora HOOK.; E: C. areola-
ta CHEESEM.; F: C. linariifolia HOOK.; G: Ventilago mad(e)ras-
patana GAERTN.

Ventilago mad(e)raspatana GAERTN. (Rhamnaceae) (C.I.Natural Orange 1)

Indigenous in the west and south parts of India, in Burma, Ceylon,
and Java.
    In the past, the root bark was traded in the form of dark red
or brown chips with a dye content of 8-10% under the names "pitti",
"raktapitta" or "pappali" as a natural dye (6). The root bark was
used as mordant dye for cotton, wool and silk, and produced a purple-
red, bordeaux-red, brown-purple, or gray to black shade, depending
on the type of mordant used (60).

Constituents (See also table III): The root bark of Ventilago
mad(e)raspatana contains ventilagin (red rosin), $C_{15}H_{14}O_6$, and the
dyes physcionanthranol A (XVI) and B (XVII), physcion (XV), and emodin (XXIV) (60).

DYES FROM DYE INSECTS

The wide varity of anthraquinone dyes identified in various madder
plants is not to be found in the dye insects American cochineal,
kermes, Polish and Armenian cochineal. These insects contain two
dyes whose constitutions are known and a third dye whose constitution has not yet been clarified. However, the fifth in the class of
dye insects, lac dye, contains five water-soluble dyes (= laccaic
acids) and traces of three water-insoluble dyes.
   Table IV illustrates the constitutions of the insect dyes,
while table V lists the dye compositions of the five dye insects
mentioned above.

Table IV. Constitutions of the red insect dyes

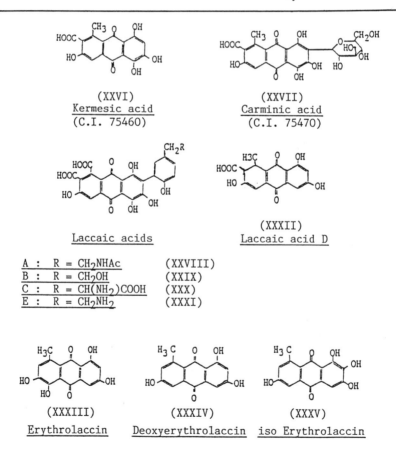

(XXVI)
Kermesic acid
(C.I. 75460)

(XXVII)
Carminic acid
(C.I. 75470)

Laccaic acids

(XXXII)
Laccaic acid D

A : R = CH_2NHAc (XXVIII)
B : R = CH_2OH (XXIX)
C : R = $CH(NH_2)COOH$ (XXX)
E : R = CH_2NH_2 (XXXI)

(XXXIII)
Erythrolaccin

(XXXIV)
Deoxyerythrolaccin

(XXXV)
iso Erythrolaccin

Table V. Hydroxyanthraquinone dyes of the dye insects

	A	B	C	D	E
Kermesic acid (XXVI) (62)	−	+	+	(+)	−
Carminic acid (XXVII) (63)	+	−	+	+	−
Flavokermesic acid* (61)	−	+	+	−	−
Laccaic acid A (XXVIII) (64)	−	−	−	−	+
Laccaic acid B (XXIX) (65)	−	−	−	−	+
Laccaic acid C (XXX) (66)	−	−	−	−	+
Laccaic acid D (XXXII) (67)	−	−	−	−	+
Laccaic acid E (XXXI) (66)	−	−	−	−	+
Erythrolaccin (XXXIII) (68)	−	−	−	−	+
Deoxyerythrolaccin (XXXIV) (68)	−	−	−	−	+
iso Erythrolaccin (XXV) (68)	−	−	−	−	+

* Constitution not yet known; gross composition: $C_{13}H_8O_6$ (61)
+ Dye detected
− Dye not detected
(+) Detection of the dye not yet definite

A : American cochineal (Dactylopius coccus COSTA)

B : Kermes (Kermococcus vermilio PLANCHON)

C : Polish cochineal (Porphyrophora polonica L.)

D : Armenian cochineal (Porphyrophora hameli BRANDT)

E : Lac dye (Kerria lacca KERR.)

Dactylopius coccus COSTA (Coccus cacti L.) (Coccidae)
(C.I.Natural Red 4)

American cochineal. Insect (female) living on the host plant Nopa-
lea coccinellifera (L.) SALM-DYCK (torch- or fig-thistle, "nopal
plant"), besides on Opuntia monacantha O.TUNA, O.vulgaris MILL. non
auct. mult., and Pe(i)reskia aculeata MILL.
Indigenous in Mexico, Central and South America. Cultivated in the
west and east parts of India, in the Canaries, in South Africa, Al-
geria, and in Spain.
    Constituents: (See also table V): American cochineal contains
up to 14% of dye consisting (exclusively) of carminic acid (XXVII).
It has been reported that cochineal contains a second dye known as
neocarminic acid (69), but this has not been confirmed. Recently, a
small amount of a second dye has been found in cochineal by HPLC (7).
    American cochineal has been identified on many old Peruvian fabrics (3, 3a, 48).

Kermococcus vermilio PLANCHON (Kermes vermilio (PLANCH.) TARG.);
formerly Kermes ilicis L. (10) (Coccidae) (C.I. Natural Red 3)

Kermes. Insect (female) living on the host plant kermes or scarlet
oak (Quercus coccifera L.)

Indigenous in the Mediterranean area and in Asia Minor.
Constituents (See also table V): Kermes contains kermesic acid
(XXVI) and a small amount of flavokermesic acid (61) whose constitu-
tion has not yet been determined.
Kermes belongs to the oldest red textile dyes (70), and it was
traded by Phoenician merchants as early as 1500 B.C. (10).

Porphyrophora polonica L. (Margarodes polonicus) (Coccidae) (C.I.Na-
tural Red 3)

Polish cochineal. Insect (female) found on roots of the knawel
(Scleranthus perennis L.) as host plant.
The host plant is indigenous on sandy soil in East Germany, Po-
land, the Ukraine, Asia Minor, the Caucasus and Turkestan.
Constituents (See also table V): Polish cochineal contains carminic acid (XXVII) and kermesic acid (XXVI), whose quantitative ratio has been determined by HPLC and is reported by Whiting to be 5:1
to 10:1 and by Wouters 15:1 (10).
Proof for the earliest use of Polish cochineal for textile dyeing was presented by Pfister when he investigated textiles that had
been found in Egypt and Syria (Palmyra) and dated from the Hellenis-
tic-Roman era. The Egyptian material originated from Persia (Dynasty
of the Sassanides, 226-641 A.D.) (70, 71), while the textile material
from the necropolis in Palmyra was found to be Chinese silk (41).
Pfister was able to distinguish by chemical analysis between red dyeings produced with kermes or with Polish cochineal.

Porphyrophora hameli BRANDT (P. armeniaca BURMEISTER)(Coccidae)

Armenian cochineal. Insect (female) that is found on the roots and
lower parts of the blades of a number of grass species belonging bo-
tanically to Aeluropus litteralis.
The host plants suitable for Armenian cochineal grow only in Ar-
menia and Azerbaijan, particularly on the soil of flat valleys, e.g.
along the banks of the Araxes river up to the north and east of Mount
Ararat (10).
Constituents (See also table V): Armenian cochineal contains
carminic acid (XXVII), as reported by Kurdian (73) in 1941 without
any indication of the identification method used. It could also contain a very small amount of kermesic acid (XXVI), as Wouters found by
investigations with HPLC (74).
Armenian cochineal may have been an important article of com-
merce in earlier times (10). Masschelein-Kleiner and Maes (75) identified carminic acid on ten samples of Egyptian textiles from the 5th
to the 7th century A.D., but not kermesic acid. This result applies
only to Armenian cochineal and not to any other of the dyes known in
antiquity. The authors also found Armenian cochineal on various Nubi-
an and Hebrew textiles.

Kerria lacca KERR (Coccus laccae, Laccifer lacca KERR) (Coccidae)
(C.I. Natural Red 25)

Lac Insect. Various species of lac insect (Lakshadia (Tachardia,
Laccifer) spp.) can be found, widely distributed, in South and South-

east Asia, for example in India, Cambodia, Thailand, Sumatra, and the Moluccas.

The host plants are, for example, Butea frondosa ROXB., Ficus religiosa L., Ziziphus jujuba MILL.

The female lac insect secretes on the twigs of the host plant a resinous substance, stick-lac, from which shellac and lac dye are produced.

Constituents (See also tables IV and V):Besides shellac rosin, lac dye contains the water-soluble dyes laccaic acid A, B, C, D, and E and three water-insoluble dyes (62-68).

In China, silk for export to west Asia was dyed with lac dye as early as in the Han dynasty (206 B.C. - 220 A.D.) (41). The Greek physician Ctesias, who lived about 400 B.C. at the Court of a Persian king, writes in his works on India known as "Indica": "There are in India insects of the size of a beetle, of the color of minium,.... They are found on trees which bear amber, ... The Indians, by bruising these animals, obtain a dye, with which they dye robes and tunics, and other articles of dress, of a scarlet color, very superior to the Persian dyes." This is a sure indication of lac dye (76).

THIN-LAYER CHROMATOGRAPHY OF MADDER DYES

Preparation of samples for TLC

Madder dyes belong to the group of mordant dyes. They are dyed on wool or silk that has been previously mordanted with aluminum or iron salts. The madder dyes react with these salts to form on the fiber color lakes that are water-insoluble and do not bleed even when treated with dilute ammonia. For the identification of madder dyes with the aid of TLC, however, we require a dye solution that can be applied to the thin-layer plate. At acid pH (pH 3 or lower), the dyes are liberated from the lake, a process during which the color visibly changes, and the organic constituents can be extracted.

In order to strip the dyes of madder dyeings completely from the fiber, they must be extracted in acid solution at temperatures up to roughly 100° C. Comparative solutions produced by extraction from dyer's plants such as madder root must also be prepared in the same manner at acid pH for the TLC. The anthraquinone dyes are present in the dyer's plants partly in the form of glycosides. These glycosides must be split hydrolytically with acids in order to obtain for the TLC a comparative solution that contains only free hydroxyanthraquinones and no hydroxyanthraquinone glycosides.

For stripping natural mordant dyes from dyeings, it is customary to use dilute hydrochloric acid or sulfuric acid, with or without an addition of organic solvents, for example, methanol, ethanol, or acetone. In the process, some hydroxyanthraquinone dyes containing carboxyl groups, for example pseudopurpurin (XXII) and Munjistin (VII), liberate carbon dioxide and are converted into purpurin (XXI) or purpuroxanthin (II). In preparing samples for TLC, any such changes must be avoided, because they make it very difficult to distinguish between dyes that are very similar to one another. The two changes caused by decarboxylation described above also occur with madder roots when they are dried or subsequently stored in drums, processes

during which an enzymatic hydrolysis of the madder glycosides and also decarboxylation are liable to occur. Madder dyes may also undergo
changes during the dyeing process itself, changes caused in most cases by the dyeing temperature, the dyeing time, and the pH of the
dyeing liquor.

In the extraction of powdered madder roots with 10% sulfuric
acid, followed by shaking with ethyl acetate, we see that the ethyl
acetate contains only a very small amount of pseudopurpurin or not at
all. However, when we investigate madder dyeings on old carpets by
the same method, the thin-layer chromatogram often shows that pseudopurpurin is present in roughly the same amount as purpurin. What is
the explanation for this ?

In a dyeing, pseudopurpurin is present, for example, as Al-Ca
lake. When boiled once with 10% sulfuric acid, this lake is split,
and the free pseudopurpurin can immediately be shaken out with ethyl
acetate before it is converted into purpurin by decarboxylation.

In the madder root, however, the pseudopurpurin is present in
the form of the relatively hydrolysis-stable glycoside galiosin. In
this case, it is necessary to heat for 20-30 minutes with dilute
acids before all dyes can be shaken out with ethyl acetate. In this
process, the pseudopurpurin is completely decarboxylated.

A search began for a new extraction method that would permit
the pseudopurpurin to be retained. The reason for this was the investigation of a pink madder lake known as "Rose madder" (C.I. Natural
Red 9). This is an artist's pigment containing mainly the aluminum
lake of pseudopurpurin. As the corresponding purpurin lake has far
lower lightfastness, lakes with a high content of pseudopurpurin are
highly desired. For investigations of this kind, it is essential to
avoid any degradation of the pseudopurpurin during the preparation
of a sample for TLC.

New improved extraction method for madder dyes.   For these investigations, the following method of extraction is recommended:
The analytical sample weighing 1-5 mg is heated in a test tube for
about one minute with 10 ml of a 1:1 mixture of 10% sulfuric acid and
ethyl acetate, butyl acetate, or toluene in a simmering water bath
until the aqueous phase is completely colorless and the organic phase
has turned yellow or orange-yellow. The lower, aqueous phase is allowed to run out of a small separating funnel, and the upper, organic
phase is shaken with water until it gives an almost neutral reaction
to pH paper. The dye solution is evaporated to dryness in vacuo or in
a porcelain dish in an air current at room temperature. The evaporation residue is taken up with a small amount of methanol or butanone-2
(for pseudopurpurin), and the solution is poured into a small 0,5-ml
or 1-ml test tube and concentrated in the test tube with an air current to obtain a sample for thin-layer chromatographic comparisons.

With this method, we have found that the sample of "Rose madder" contains only traces of purpurin besides of pseudopurpurin. The
method is also useful for investigating madder types that contain only small amounts of purpurin (formed probably from pseudopurpurin) as
well as pseudopurpurin, but no alizarin. Examples of these madder types
are the South American relbun roots such as Relbunium hypocarpium (L.)
HEMSL. and R. ciliatum (L.) HEMSL. and the Japanese madder roots of
Rubia akane.

This improved method of extraction is also useful for preparing
reference solutions of dyer's plants for TLC comparisons. In this case, however, the extraction time must be lengthened to about 5 minutes. Despite this, however, the pseudopurpurin is mostly retained, as
can be confirmed by thin-layer chromatographic comparison.

For investigating dyeings with madder dyes, the extraction time
of one minute is adequate, as described in the analysis of the "Rose
madder" sample.

TLC procedure for madder dyes

When we strip a red dyeing by the improved extraction method with a
mixture of 10% sulfuric acid and ethyl acetate (1:1) and the upper
layer has a yellow or orange-yellow shade and the lower layer is colorless, we can assume that this is a madder dyeing.

The extract of the supposed madder dyeing is applied after concentration as described above together with the following reference
solutions to a Mikropolyamid F 1700 thin-layer plate (Schleicher &
Schüll): alizarin, purpurin, pseudopurpurin, or reference solutions
obtained from the following dyer's plants by the improved extraction
method: Rubia tinctorum, R. peregrina, R. cordifolia, R. akane; Galium verum (or G. mollugo), Oldenlandia umbellata, Morinda citrifolia,
M. umbellata, Coprosma lucida, and Ventilago mad(e)raspatana. The
chromatogram is developed in a TLC separation chamber with the solvent mixture toluene-acetic acid (9:1), or sometimes butanone-2 -
formic acid (95:5), or chloroform-methanol (95:5) over a distance of
8-10 cm (2). When the chromatogram has dried, the inherent colors of
the spots, and then their UV fluorescence under the UV lamp, are compared, and any conformity between the unknown and the comparative
samples is determined. The thin-layer chromatogram is the immersed
for about 20 seconds in a 0,5% solution of uranyl acetate in 50% methanol. After this, the chromatogram is pressed between several layers of filter paper and dried in the air. The spots of the uranyl
lakes are then compared to see whether they coincide.

Preparation of an album with thin-layer chromatograms of tested
samples facilitates the evaluation.

Figures 1 and 2 show thin-layer chromatographic comparisons of the
dyes belonging to various types of madder, as listed in the table II
and III. The chromatographic conditions are the same for these two
chromatograms, viz. layer material: Mikropolyamid F 1700; solvent:
toluene-acetic acid (9:1); color reaction: uranyl acetate.

Figure 1 shows a chromatogram with the dyes of various Rubia, Galium,
and Relbunium species. The following hydroxyanthraquinones on this
chromatogram indicate clearly the presence of a definite dye plant of
the mentioned species:
1 Rubia tinctorum: alizarin and purpurin can be clearly identified
   by the marked spots. Pseudopurpurin has been converted into purpurin by decarboxylation, because the old extraction method (boiling with 10% sulfuric acid, followed by shaking with ethyl acetate) has been used to isolate the dyes from the madder roots.
   Even with the new, improved extraction method, we often obtain
   this chromatographic picture in analyzing old madder dyeings, if

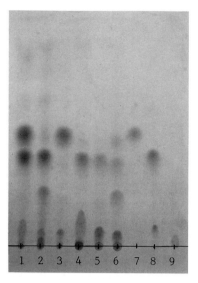

Figure 1. TLC of dyes of various madder types. 1, madder (Rubia tinctorum L.); 2, wild madder (R. peregrina L.); 3, Indian madder (R. cordifolia L.); 4, Japanese madder (R. akane); 5, relbun root (Relbunium
hypocarpium (L.) HEMSL.); 6, lady's bedstraw (Galium verum L.); 7, alizarin; 8, purpurin + pseudopurpurin (lower spot); 9, munjistin.

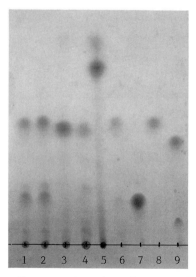

Figure 2. TLC of dyes of some other dye plant similar to the madder types; 1, suranji (Morinda citrifolia L.); 2, mang-kouda (Morinda umbellata L.); 3, chay root (Oldenlandia umbellata L.); 4, coprosma root
(Coprosma lucida J.R. & G. Forst); 5, pitti (Ventilago mad(e)raspatana
GARTN.); 6, morindon; 7, emodin; 8, alizarin; 9, purpurin + pseudopurpurin (lower spot).

the pseudopurpurin has already been destroyed during drying of the madder or in the dyeing process.

The red spot near the starting line is probably a hydroxyanthraquinone glycoside, that has not been hydrolyzed.

2   Rubia peregrina: As the new, improved extraction method was used here, the pseudopurpurin spot is most prominent, and purpurin can also be clearly identified. Alizarin is only present in traces. The red spot may be rubiadin (III), until now not mentioned in the literature as a dye in Rubia peregrina.

3   Rubia cordifolia: In this case, too, the new extraction method was used. The pronouced spot is that of alizarin, and the second is that of pseudopurpurin. The red spot of (probably) rubiadin in contrast to Rubia peregrina cannot be found.

4   Rubia akane: Extraction by the old method, but only for two minutes.the relatively stable pseudopurpurin glycoside has not been hydrolyzed by this treatment.

5   Relbunium hypocarpium: New extraction method. Contains almost exclusively pseudopurpurin and only a small amount of purpurin.

6   Galium verum: New extraction method. Pseudopurpurin is the main component, but purpurin, and in contrast to Rubia peregrina, alizarin can also be clearly identified. The red spot is, as in the case of Rubia peregrina, probably rubiadin.

Figure 2   shows a chromatogram with the dyes of Morinda, Oldenlandia, Coprosma, and Ventilago species. These dye plants are similar to the Rubia spp.

1   Morinda citrifolia and   2 M.umbellata   contain morindon (XXIII) as the principal dye and also a small amount of (probably) emodin (XXIV). It is not possible to distinguish between the two Morinda species by TLC.

3   Oldenlandia umbellata: Contains almost exclusively alizarin as dye and can thus be distinguished from all other madder types. This dyer's plant cannot be clearly distinguished from synthetic alizarin by this method.

4   Coprosma lucida: On the chromatogram are two prominent spots, viz. a violet-brown spot roughly at the level of alizarin, and a gray-brown spot near the starting line. These two spots could not be identified, because reference compounds were not available.

5   Ventilago mad(e)raspatana: The dyes from this dyer's plant differ on the thin-layer chromatogram from all other madder dyes so markedly that the plant extract can be used for identification by TLC comparison. Although the dyes of Ventilago mad(e)raspatana are not quite unknown to us (see table III), it is not possible to classify the spots, because the necessary comparative material is missing.

Figure 3 illustrates a thin-layer chromatogram in which the old and the new extraction methods for madder dyes are compared with each other. The chromatographic conditions are the same as those used for the thin-layer chromatograms in figures 1 and 2. The two extraction methods were compared to determine their usefulness for testing a sample of the artist's pigment "Rose Madder Genuine" (C.I.Natural Red 9) (see page 202), a pigment whose quality depends on its content of pseudopurpurin (XXII).

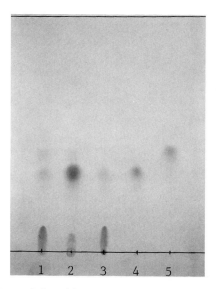

Figure 3. Comparison of the old and the new (improved) extraction methods for madder dyeings and madder pigments (see page 202) by TLC. 1, and 2, extractions of a sample of the artist's pigment "Rose Madder Genuine"; 1, extraction by the new method; 2, extraction by the old method; 3, pseudopurpurin (lower spot); 4, purpurin; 5, alizarin.

1   Extraction of the "Rose Madder Genuine" sample with a mixture of the same amounts by volume of 10% sulfuric acid and toluene, one minute in the simmering water bath (new extraction method). The chromatogram shows clearly that the sample contains almost exclusively pseudopurpurin, besides traces of purpurin and alizarin, and that any degradation by decarboxylation and conversion into purpurin is, at most, minimum.

2   Extraction of the same sample by the old method (see page 201), In this case, the thin-layer chromatogram shows purpurin as the main component; thus most of the pseudopurpurin has been destroyed.

3   Pseudopurpurin, made from a synthetic pseudopurpurin lake by extraction with the new method. In contrast to the sample of "Rose Madder Genuine" (made from madder roots) besides of the principal dye only traces of purpurin have been found, but not alizarin. Therefore is is possible to distinguish natural and synthetic pseudopurpurin lakes by this method.

Figure 4 is a photograph of an Adlerdalmatika, Chinese, about 1300 A.D. This Adlerdalmatika is one of the coronation robes used by the emperors of the Holy Roman Empire of the German Nation, which are exhibited in the Weltliche Schatzkammer of the Vienna Imperial Palace. During the restoration of this robe in the Kunsthistorische Museum, Sammlung für Plastik und Kunstgewerbe, in Vienna 1986, the dyes on the ground fabric were investigated. A sample weighing 80 mg was available for the analysis.
Result: The warp and filling threads are composed of natural silk, grounde with the lichen dye orchil (C.I.Natural Red 28) and dyed with madder (Rubia tintorum).

Figure 4. Aldlerdalmatika, Chinese, about 1300 A.D. (Reprinted with permission. Copyright 1987 Kunsthistorisches Museum Wien.)

Analytical procedure: The warp and filling threads were first separated so that they could be tested separately. With the improved extraction method (10% sulfuric acid + ethyl acetate (1:1)), the madder was
repeatedly stripped from 5 mg samples of the warp and the filling
threads until the ethyl acetate was no longer stained yellow. After
this extraction, the stripped silk sampled remained red. Pure madder
dyeings are only slightly yellow after this treatment. Consequently,
there must be a second red dye on the silk. This dye remains red after
the acid treatment, and turns violet after washing-out with water and
addition of ammonia. When sodium dithionite is also added, the dyeing
becomes colorless when the solution is boiled. After pouring the sodium
dithionite vat off and washing the silk sample, the violet shade re-
turns on exposure to the open air, and it turns red after evaporation
of the ammonia residue. This is, according to my own experience, a specific proof of orchil dyeings (2).
        The extracts with 10% sulfuric acid + ethyl acetate are concen-
trated by the method described on page 15 to obtain a suitable sample
for the detection of madder by TLC comparison, as illustrated in figure 5.

Figure 5 shows the comparison of the madder dyes taken from the Adlerdalmatika with samples of Rubia tinctorum, alizarin, and purpurin by
TLC. The dye extracted from the sample of the Adlerdalmatika with dilute sulfuric acid is unmistakably madder.
        What can we conclude from the result of the dye investigation
carried out on the red sample from the Adlerdalmatika ?

There are indications that lichen dyes such as orchil have been used
for grounding madder dyeings (53). It is likely that even the Phoeni-
cians used lichen dyes for grounding textiles dyed with Tyrian Purple
to cut down the amount of this costly dye in dyeing deep shades (77).
Purple dyeings grounded with orchil were known as "conchoid purples"
(78). It is also possible that the purpose of grounding the silk of
the Adlerdalmatika with orchil was also to weight the silk.
        As orchil dyeings have very poor lightfastness, I recommended
suitable protective measures for storing the Adlerdalmatika.

THIN-LAYER CHROMATOGRAPHY OF INSECT DYES

Preparation of samples for TLC

When the improved method of extraction is used for stripping madder
dyes with 10% sulfuric acid and ethyl acetate (see page 202), and the
ethyl acetate layer is not stained yellow, but orange, this could indicate the presence of insect dyes containing kermesic acid (XXVI).
If the sulfuric acid layer is stained orange to red, this could indicate the presence of carminic acid (XXVII) or laccaic acids (XXVIII –
XXXII). If the sulfuric acid layer is colorless and the ethyl acetate
layer is orange, the ethyl acetate layer is separated off in a small
separating funnel and a specimen for the TLC is prepared as described
above for madder dyeings (page 202).
        If the sulfuric acid layer is stained orange to red after treat-
ment with 10% sulfuric acid and ethyl acetate, 3-methylbutanol-1 is
added in roughly the same amount as ethyl acetate, and the solution
is shaken vigorously. After the phase separation in the separating

funnel, the lower layer is colorless and the upper layer is orange
to red. The lower sulfuric acid layer is discarded, and a sample for
the TLC is prepared from the upper phase, as described above for madder dyeings (page 202).

TLC procedure for insect dyes

The solutions obtained after concentration of the ethyl acetate and/
or ethyl acetate - 3-methylbutanol-1 solution(s) are applied side by
side with the following reference solutions to a Mikropolyamid F 1700
thin-layer plate: American cochineal, kermes, Polish cochineal, and
laccaic acids from lac dye.
       The following solvents are suitable for separating insect dyes:
butanone-2 - formic acid (7:3)   (2)
iso propanol - formic acid - water (7:1:3) (79)
       For the separation of the kermes dyes kermesic acid and flavo-
kermesic acid the following solvents are suitable according to my own
experience:
methanol - formic acid - water (8:1:1)
iso propanol - formic acid - water (6:2:2)
       The chromatographic separation and the identification with ura-
nyl acetate are carried out as in the TLC of madder dyes.

Figure 6 illustrates the following four dye insects in the dried form
that were used in the past for dyeing textile materials:
Dactylopius coccus COSTA
Kermes vermilio PLANCH. (TARG.) (Revised name: Kermococcus vermilio
Porphyrophora hameli BRANDT   PLANCHON)
Margarodes polonicus   (Revised name: Porphyrophora polonica L.)
As the illustration shows, the polish cochineal (Margarodes polonicus)
is far smaller than the other three insects, which are all more or less
of the same size.

Figure 7 and figure 8 illustrate two fragments of Karabagh carpets
from the 19th century containing a red that has probably been dyed with
Armenian cochineal. In various carpet books (80,81) it is pointed out
that "Armenian red" is also known as "Karabagh red", because it was
used most frequently and for the longest time in the Armenian enclave
Nagorny-Karabakh in Azerbaijan.

Figure 7 illustrates the fragment of a Karabagh carpet, first half of
the 19th century, that I received from the owner Detlef Lehmann, Textile Restorer, D-2943 Esens (Federal Republic of Germany). This carpet
fragment has red knots dyed with a dye containing carminic acid as the
principal component and traces of other hydroxyanthraquinone dyes. It
is possible that this is a dyeing with Armenian cochineal. As the synthetic dye Magenta (C.I.42510), which was only found in 1859, was also
identified in a faded red-violet in this carpet, the date claimed for
this carpet, viz. first half of the 19th century, is not quite correct.

Figure 8 illustrates the fragment of another Karabagh carpet from the
middle of the 19th century that I received also from D.Lehmann under

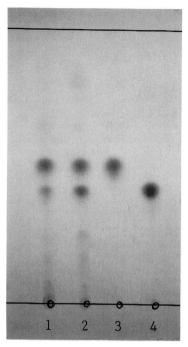

Figure 5. Identification of the madder dyes in the Aldlerdalmatika by TLC comparison. 1, dyes extracted from the Aldlerdalmatika by diluted sulfuric acid; 2, madder dyes; 3, alizarin; 4, purpurin.

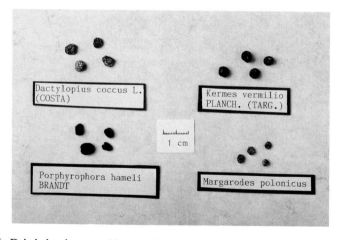

Figure 6: Dried dye insects. Above left, American cochineal; above right, kermes; below left, Armenian cochineal; below right, Polish cochineal.

silk thread from this valuable coronation robe from Dr. Rotraud Bauer (Kunsthistorisches Museum, Vienna). With the aid of color reactions, he in his laboratory and I in mine proved that it was not a dyeing with Tyrian purple, as assumed in the past, but a dyeing with an insect dye.

This coronation robe for the emperors of the Holy Roman Empire of the German Nation was made in 1133–34 for the Norman king Roger II in the royal court workshop in Palermo. In 1194, after the Normans had been driven out of Palermo, it passed into the possession of the Hohenstaufen together with the treasure of the Normans. When the coronation robe was being restored in 1986, I received from Dr. R. Bauer a sample of the red silk (22,5 mg) and samples of the warp (2,9 mg) and filling threads (17,4 mg) of the lining of this robe with the request to carry out a dye analysis.

Another robe belonging to the coronation costumes of the emperors of the Holy Roman Empire of the German Nation is the Tunicella illustrated in Figure 11 , which was made in the first half of the 12th century in the royal court workshop in Sicily. I received 23,9 mg of the dark blue dress material and 15,3 mg of the red border for a dye analysis. IR spectra comparison of the dye shaken out of the vat with ethyl acetate showed unambiguously that indigo had been used to dye the dark blue dress material.

Figure 12 shows the identification of the three red samples of the coronation robe and of the red sample of the border of the Tunicella by thin–layer chromatographic comparison on Mikropolyamid F 1700 as

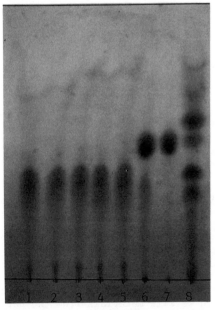

Figure 12. Identification of the insect dyes in the Sicilian coronation robe (figure 10) and in the Tunicella (figure 11) by TLC comparison. Solvent: butanone-2 - formic acid (7:3); 1-3, extracts of the three samples of dyeings of the coronation robe; 1, red silk; 2, lining material, filling threads; 3, lining material, warp threads; 4, red silk from the border material of the Tunicella; 5, kermes; 6, Polish cochineal; 7, American cochineal; 8, laccaic acids from lac dye.

layer material. Solvent is the mixture: butanone-2 - formic acid
(7:3). Color reaction: Uranyl acetate.
   The thin-layer chromatogram shows at first sight that the red
samples of the coronation robe and of the Tunicella have been dyed
with kermes (Kermococcus vermilio PLANCHON). As the comparison shows,
not only the green spot of kermesic acid, but also the red spot of
flavokermesic acid is clearly visible.

Acknowledgements

I wish to thank Max Saltzman, Los Angeles, for his untiring efforts
to obtain extremely rare comparison materials and for the unselfish
advice that he always gave me in overcoming analytical problems in
the field of natural dyes.
   Acknowledgements are also due to Dr. Rotraud Bauer, Vienna,
for providing me with large analytical samples of the coronation
robes of the emperors of the Holy Roman Empire of the German Nation.
Without her assistance, it would not have been possible for me to
present such precise analytical results.

Literature Cited

1. Whiting, M.C., "A Report on the Dyes of the Pazyryk Carpet",
   Oriental Carpet & Textile Studies. Pinner, R., Denny, W.B., Ed.;
   Vol. I; Published in Association with HALI Magazine, London,
   1978, 18-22.
2. Schweppe, H. In "Historic Textile and Paper Materials. Conservation and Characterization"; Needles, H.L., Zeronian, S.H.,
   Ed.; ADVANCES IN CHEMISTRY SERIES No.212; American Chemical Society: Washington, D.C., 1986, pp. 153-74.
3. Saltzman, M. "Analysis of Dyes in Museum Textiles" or "You Can't
   Tell a Dye by Its Color", in: Textile Conservation Symposium in
   Honor of Pat Reeves; McLean, C.C., Connell, P., Ed.; The Conservation Center, Los Angeles County Museum, Los Angeles, Calif.,
   1986, pp. 27-39.
3a. Saltzman, M. In "Archaeological Chemistry II"; Carter, G.F., Ed.;
   ADVANCES IN CHEMISTRY SERIES No. 171; American Chemical Society:
   Washington, D.C. 1978, pp.173-85.
4. Schweppe, H., Die BASF (Aus der Arbeit der BASF Aktiengesellschaft) 1976, 26, 29-36.
5. Ford, P.R.J, "Das Rätsel der Polenteppich-Farben", in: Heimtex
   (Fachzeitschrift für die gesamte Innenraumausstattung), Herford
   (FRG), 1986/1, pp. 143-55; 1986/2, pp. 48-55.
6. "Colour Index 3rd ed."; The Society of Dyers & Colourists: Brad-
   fors, United Kingdom, 1971.
7. Wouters,J., Stud. Conserv. 1985, 30, 119-28.
8. Taylor, G.W., Text. History 1987, 18(2), 143-6.
9. Hofenk-de Graaff, J.H.; Roelofs, W.G.H., "On the occurence of
   red dyestuffs in textile materials from the period 1450-1600";
   ICOM Plenary Meeting; Madrid, Oct. 1972.
10. Donkin. R.H., Anthropos, Freiburg (FRG), 1977, 72, 847-80.
11. Taylor, G.W., "Survey of the Insect Red Dyes", in: Dyes on Historical and Archaeological Textiles; 3rd Meeting, York Archaeological Trust, Sept. 1984, 22-5.

12. Whiting, M.C., Chemie in unserer Zeit 1981, 15, 179–86.
13. Mushak, P.; O'Bannon, G.W., Orient. Rug Rev. 1982, Vol.II, No.10, 6–8.
14. Ibid 1983, Vol. III, No.2, 16–9.
15. Masschelein-Kleiner, L., Microchim.Acta, 1967, 6, 1080–5.
16. Masschelein-Kleiner, L.; Heylen,J.B., Stud.Conserv. 1968, 13, 87–97.
17. Masschelein-Kleiner, L.; Maes, L., Bull. Inst.R Patrimoine Art 1970, t. XII, 34–41.
18. Masschelein-Kleiner, L.; Znamensky-Festraets, N.; Maes, L., Bull. Inst. R Patrimoine Art 1969, T. XI, 269–72.
19. Roelofs, W.G.H., "Thin-layer chromatography, an aid for the analysis of binding materials and natural dyestuffs from the work of art"; ICOM Plenary Meeting; Madrid, Oct. 1972.
20. Schweppe, H., Z. anal. Chem. 1975, 276, 291–6.
21. Schweppe, H., HALI (Int. J. Orient. Carpets Text.) 1979, Vol. II, No.1, 24–7.
22. Schweppe, H., J. Am. Inst. Conserv. Hist. Art. Works 1980, 19, 24–7.
23. Böhmer, H., HALI (Int. J. Orient. Carpets Text.) 1979, Vol. II, No. 1, 30–3.
24. Brüggemann, W.; Böhmer, H., "Teppiche der Bauern und Nomaden in Anatolien", Verlag Kunst und Antiquitäten GmbH: Hannover (FRG), 1980, pp. 88–118.
    English Edition: Brüggemann, W.; Böhmer, H., "Rugs of the Peasants and Nomads of Anatolia", published by Kunst und Antiquitäten, Munich (FRG).
25. Mushak, P., Orient. Rug. Rev. 1983, Vol. III, No.4, 3–5.
26. Whiting, M.C., HALI (Int. J. Orient. Carpets Text. 1978, Vol I. No.1, 39–43.
27. Ibid 1979, Vol.II, No. 2, 28–9.
28. Kharbade, B.V.; Agraval, O.P., J. Chromatog. 1985, 347, 447–54.
29. Airaudo, CH.B.; Cerri, V.; Gayte-Sorbier, A.; Andrianjafiniony, J., J. Chromatog. 1983, 261, 272–85.
30. Daniels, V., "Progess in the dye analysis in the British Museum", in: Dyes on historical and archaeological Textiles; 3rd Meeting, York Archaeological Trust, Sept.1984, 8.
31. Harvey, J., "Analysis of dyes in fabrics recovered from the Mary Rose Site", in: Dyes on Historical and Archaeological Textiles; 1st Meeting, York Archaeological Trust, Aug. 1982, 3.
32. Burnett, A.R.; Thomson, R.H., J. chem. Soc. (C) 1968, 2438–41.
33. Berg, W.; Hesse, A.; Herrmann, M.; Kraft, R., Pharmazie 1975, 30(5), 330–4.
34. Hager's Handbuch der pharmazeutischen Praxis. 4.Neuausgabe, 6. Band: Chemikalien und Drogen, Teil B (R-S), 179–83. Berlin, Springer-Verlag 1979.
35. Vaidyanathan, A., Dyes & Pigments 1985, 6, 27–30.
36. Pfister, R., "Les Textiles du Tombeau de Toutenkhamon", Revue des Arts Asiatiques 1937, 11, 207–18.
37. Gulati, A.N.; Turner, A.J., "A Note on Early History of Cotton", in: J. of the Text. Inst. 1929, 20, T1–T9.
38. Murty et al., Ind. J. Chem. 1970, 8, 779.
39. Berg, W. et al., Pharmazie 1974, 29, 478.
40. Tessier, A.M.; Delaveau, B.; Champion, B., Planta Med. 1981, 41, 337–43.

41. Pfister, R., Nouveaux Textiles de Palmyre, Vol. I-III, Paris, Les Éditions d'Art et d'Histoire, 1934-1937-1940.
42. Irwin, J.; Hall, M., Indian painted and printed fabrics, Ahmedabad, India, 1971.
43. Wada, M., Kagaku (Science) 1941, 11, 416 (In Japanese).
44. Hayashi, K., "Chemical procedure for the Determination of Plant Dyes in Ancient Japanese Textiles", in: Proceedings 2nd ISCRCP. Cultural and Analytical Chemistry 1979, 39-50.
45. Feller, R. L.; Curran, M.; Bailie, C., "Identification of Traditional Organic Colorants employed in Japanese Prints and Determination of their Rates of Fading", in: "Japanese Wood Prints: A Catalogue of the Mary Ainsworth Collection". (Allen Memorial Art Museum, Oberlin College, Oberlin, Ohio. Distributed by Indiana University Press, 1984.
46. Schweppe, H., Unpublished tests.
47. Hayashi, K.; Suzushino, G., Sci. Papers on Jap. Antiq. and Art Crafts 1951, 3, 40-4 (in Japanese).
48. Fester, G.A., Isis 1953, 44, 13-6.
49. Fester, G.A.; Lexow, S.G., Revista de la Facultad de Quimica Industrial y Agricola, Santa Fe, 1942/43, 11/12, 84/ 112.
50. Klein, O., "Zur Geschichte der Araukaner", Ciba-Rundsch. 1961/ 62, 2-25.
51. Burnett, A.R.; Thomson, R.H., J. chem. Soc. (C) 168, 854-7.
52. Grierson, S., J. Soc. Dyers & Col. 1984, 100, 209-11.
53. Telfer Dunbar, J., Ciba-Rundsch. 1951, 98, 3605-8.
54. Schweppe, H., Result of an analysis, given by letter to: Schleswig-Holsteinisches Landesmuseum, Schleswig, Federal Republic of Germany.
55. Vogler, H., "Arbeitsmethoden und Farbstoffe der altindischen Färber", in: Deutscher Färberkalender 1982, 209-32.
56. Hager's Handbuch der pharmazeutischen Praxis. 4. Neuausgabe, 5. Band: Chemikalien und Drogen (H-M), 892-3. Berlin, Springer-Verlag, 1976.
57. Burnett, A.R.; Thomson, Phytochemistry 1968, 7, 1421-2.
58. Briggs, L.H. et al., J. chem. Soc. 1965, 2595.
59. Hager's Handbuch der pharmazeutischen Praxis. 4.Neuausgabe, 4. Band: Chemikalien und Drogen (CI-G), 293-4. Berlin, Springer-Verlag, 1973.
60. Hager's Handbuch der pharmazeutischen Praxis. 4. Neuausgabe, 6. Band: Chemikalien und Drogen, Teil C (T-Z), 399-400. Berlin, Springer-Verlag, 1979.
61. Dimroth, O.; Scheurer, W., Liebig's Ann. 1913, 399, 43-61.
62. Gadgil, D.D.; Rama Rao, A.V.; Venkataraman, K., Tetrahedron Lett. 1968, 2223-7.
63. Bhatia, S.B.; Venkataraman, K., Indian J. Chem. 1965, 3, 92-3.
64. Pandhare, A.V.;.Rama Rao, A.V.; Shaikh, I.N., Indian J. Chem. 1969, 7, 977-86.
65. Bhide, N.S.; Pandhare, E.D.; Rama Rao, A.V.; Shaikh, I.N.; Srinavasan, R., Indian J. Chem. 1969, 7, 987-95.
66. Rama Rao, A.V.; Shaikh, I.N.; Venkataraman, K., Indian J. Chem. 1969, 7, 188-9.
67. Mehandale, A.R.; Rama Rao, A.V.; Shaikh, I.N.; Venkataraman, K., Tetrahedron Lett. 1968, 2231-4.

68. Mehandale, A.R.; Rama Rao, A.V.; Venkataraman, K., Indian J. Chem. 1972, 10, 1041-6.
69. Frey, 1931, Zur Kenntnis des Karmins und der Neokarminsäure. Dissertation Technische Hochschule Zürich.
70. Lillie, R.D., J. Soc. Dyers & Col. 1979, 95, 57-61.
71. Pfister, R., La Décoration des étoffes d'Antinoe, in: Revue des Arts Asiatiques 1928, 5, 215-43.
72. Pfister, R., Etudes textiles, in: Revue des Arts Asiatiques 1934, 8, 77-92.
73. Kurdian, H.,"Kirmiz", in:J. Amer. Oriental Studies 1941, LXI, 105-7.
74. Wouters, J., Information by letter to Max Saltzman, Los Angeles (Nov 12, 1987).
75. Masschelein-Kleiner, L.; Maes, L., "Ancient Dyeing Techniques in Eastern Mediterranean Regions", ICOM Zagreb 1978, paper 78/9/3.
76. Fleming, H., Textile Colorist 1930, 52, 696-703.
77. Perkins, P., J. Soc. Dyers & Col. 1986, 102, 221-7.
78. Born, W., Ciba-Review 1937, 4, 113.
79. Kanda, H., Jap. J. Food Sanitation Research (Shokuhin, Eisei, Kenkyu) 1985, 35, 813-20; (Ref.: CAMAG-Literaturdienst).
80. Hubel, R.H.,"Ullstein Teppichbuch", Frankfurt,Ullstein 1974,p.40
81. Biedrzynski, E., "Bruckman's Teppich-Lexikon", Munich, 1975, p.125.

RECEIVED February 28, 1989

Chapter 14

Ultraviolet and Infrared Analyses of Artificially Aged Cellophane Film

Laura DeSimone and Ira Block

Department of Textiles and Consumer Economics, University of Maryland,
College Park, MD 20740

The thermal oxidation of unplasticized
Cellophane film at temperatures from 90 to 140 °C
was evaluated by the increase in absorbance at 265
nm and the decrease in percent transmission at 1725
cm^{-1}. Treatment of the aged films with aqueous
$NaBH_4$ solution almost completely eliminates the
peaks, indicating that the absorption in both
regions is due almost exclusively to carbonyl
moieties. A water wash partially decreases the
absorbance, showing the presence of low-molecular-
weight species.
     The oxidation of the film follows first-order
kinetics with rates close to those found for decreases in tear strength and depolymerization of
cotton and rayon fabrics. The reaction mechanisms
appear not to be affected by temperature. The infrared spectra of the film and of the water extract
of the aged film are essentially the same as those
found for naturally aged linen.

Cellulose consists of a chain of beta-D-glucopyranose residues linked
together at the C1 and C4 positions by beta-glycosidic bonds (1).
The hydroxyl groups at C2, C3 and C6 may undergo the usual reactions
of secondary and primary alcohols, respectively. The reactivity of
these groups depends upon their position and the conditions and
agents of the reaction (1). Atmospheric oxygen is a major factor in
the deterioration of cellulose (2). In the presence of oxygen, cellulose is converted to a product which contains carboxyl, carbonyl
and peroxide groups (3,4). As a result of the formation of these new
species, the oxidized cellulose exhibits changes in behavior and
properties.
     Oxygen is a non-specific oxidant: that is, no specific position
along the cellulose chain is preferentially attacked (2,5). However,

0097–6156/89/0410–0220$06.00/0

the more accessible regions of the material will be converted before the more crystalline material undergoes reaction (2,5,6). Further oxidation may be terminated or the reaction may slowly proceed into the more highly crystalline areas of the material. Extreme oxidation of cellulose by a non-specific agent results in a brittle, highly crystalline product (7).

The oxidation of cellulose results in the formation of carbonyl (aldehyde and ketone), and/or carboxyl groups. The formation of carbonyl groups may destabilize the glycosidic link (8), cause direct ring scission, or promote a beta-alkoxy elimination reaction with subsequent ring or glycosidic bond cleavage (7). Carbonyl groups may oxidize further to carboxyl groups; the presence of which may promote acid hydrolysis of the acetal link.

In addition, when the chain ends in an aldehyde, the cellulose may be "unzipped" with the sequential removal of one glucose unit at a time. In this manner, low-molecular-weight fractions are produced. The mechanism is more fully described in References 1 and 9.

A fuller understanding of the mechanisms by which cellulose is degraded upon aging is, of course, of major importance to conservators. If the major process is through acid hydrolysis, then appropriate treatments would take the form of deacidification and alkalization. If, on the other hand, carbonyl species cause the most damage, reduction treatments would be more suitable. It may also be appropriate to combine both reduction and alkalization treatments (10).

Cellophane film is prepared from regenerated cellulose and is similar to rayon fiber in that it has a lower molecular weight than cotton and contains a small amount of hemicellulose, as does linen. Cellophane film, therefore, although not a duplicate of any natural fiber, is similar enough in chemical structure and morphology to make it useful as a model system. Moreover, its transparency and the precision of its manufacture make it quite useful for this type of study.

EXPERIMENTAL

Materials. Unplasticized Cellophane film was provided by E. I. duPont de Nemours & Co., Wilmington, DE, and was used without further treatment. The variation in thickness was small enough to be of little concern in quantitative measurements of absorbance and percent transmission.

Sodium borohydride ($NaBH_4$), 98% pure, was obtained from Alfa Products, Inc, Danvers, Ma and used as obtained.

Accelerated Aging. The Cellophane samples were artificially aged in a Thelco forced-draft oven at 90, 110, and 140 °C. The relative humidity varied, but was less than 2% at all temperatures. Samples were sandwiched between a glass fiber screen and glass microscope slides to prevent curling. After removal from the oven, samples were

placed in plastic bags and stored in desiccators over Drierite for at
least 48 hours before testing.

Spectrophotometry. Infrared spectra were obtained with a Perkin
Elmer Model 281B Spectrophotometer interfaced with a computer data
station. Films were encased in sample holders in the reference beam,
and spectra were obtained with air as reference. The spectra were
scanned from 1900 - 1500 cm^{-1} at a scanning time of 60 minutes with
a response setting of 2. The slit program was set at N. All spectra
were recorded on computer disks.

A Beckman Model 25 UV-Visible Spectrophotometer was used to
determine the ultraviolet spectra of the Cellophane samples. Sample
and reference films were held in custom-made sample holders. Scans
of the UV-Visible spectrum indicated an absorption peak at about 265
nm. Therefore, all further measurements were made at the wavelength
of maximum absorption in the 255 - 270 nm region.

It was noted that films stored in desiccators absorbed water very
rapidly. This phenomenon was exhibited in curling of the films and
changes in the IR spectra in the 1600 cm^{-1} region during the first few
minutes. Therefore, spectrometry was performed after films had been
conditioned for one hour.

Borohydride Treatment. A 0.05 M solution was prepared by the addition of 1.9 g of solid to sufficient deionized water to dissolve the
NaBH$_4$ powder. The total volume was then adjusted to 1000 ml. Films
of approximately 0.725 g were soaked in 100 ml of the solution for
one hour. The samples were removed from the solution, rinsed in
deionized water and blotted. In order to prevent wrinkling of the
film, the samples were placed between dry filter papers, and weighted
until dry. Concurrently, control samples that had been soaked for
one hour in deionized water, were dried in the same manner. After
air drying, samples were stored in desiccators until needed.

Extracts. Extracts from water-washed film samples were collected and
analyzed. Both unbaked and baked film samples were weighed, soaked
in one hundred ml of water, rinsed and dried. The water from the
soaking was reserved and the pH was measured with an Instrumentation
Laboratories Mod. 265 pH/mV meter. All extracts were found to be
mildly acidic, with the samples subject to longest baking times having the lowest pH. Each liquid extract was examined in the Beckman
UV spectrophotometer using matched cuvettes to compare the sample to
a deionized water reference. UV spectra were recorded from 330 to
200 nm.

In addition, aqueous extracts, obtained as above, were allowed
to evaporate to dryness at room temperature. The remaining residue
was added to approximately 500 mg of KBr and was made into a disk for
infrared analysis. The spectra were recorded in the same manner as
for the films.

Kinetics. It has been generally accepted that the overall degradation of cellulosic materials is adequately described by a first-order kinetic model ($\underline{11 - 14}$). Thus, if oxidized cellulose (B) is formed directly from cellulose (A) and oxygen (O_2) by the reaction

$$A + O_2 \quad > B,$$

one may write the rate equation as follows, where k' is the reaction rate constant:

$$d[A]/dt = k'* [A][O]. \tag{1}$$

Since the oxygen concentration remains constant, Equation 1 may be rewritten as

$$d[A]/dt = k * [A]. \tag{2}$$

Solving for [A] yields

$$A = A_0 * \exp(-kt). \tag{3}$$

Since $[B] = A_0 - [A]$, the concentration of the oxidized material is

$$B = A_0 * [1- \exp(-kt)], \tag{4}$$

which is also the amount of A which has reacted. This latter parameter is usually given the symbol "X". Since $[A] = A_0 - X$, Equation 4 may be rewritten as

$$Ln[(A_0/(A_0 - X)] = k * t. \tag{5}$$

Thus, Equation 5 can be substituted for the unwieldy exponential Equation 4 so that linear regression may be performed on the plotted data, and rates may be determined and compared.

In the following graphs, a curve of the form of Equation 4 was fitted to the data by use of Lotus-123 on an IBM PC/XT. The curve was adjusted for best fit by plotting the calculated values versus the data and adjusting the calculation until the best linear regression was obtained. Using this technique on the 140 °C data yielded the value for A_0. The data were then plotted in the form of Equation 5. It was found that the values for UV absorbance were about 10x those for IR absorbance. Thus, the UV measurements yield more precise results than do the IR data. The following discussion is based on the UV measurements.

RESULTS AND DISCUSSION

Spectroscopy

The IR spectra of as-received and aged Cellophane film are shown in Figure 1. The absorption maxima (transmission minima) in the 2800 −

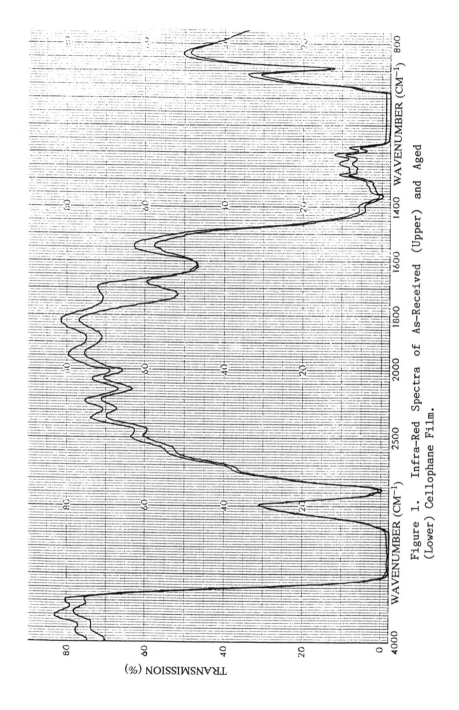

Figure 1.   Infra-Red Spectra of As-Received (Upper) and Aged (Lower) Cellophane Film.

3700 cm^{-1} region are due to the stretching vibrations of –OH and –CH– moieties (15). The shoulder at 2700 – 2800 may be caused by the – CH– stretching of aldehydes (16). Peaks in the region from 1800 to 2700 cm^{-1} are an interference pattern produced by light reflected from the front and back surfaces of the film. Absorption in the 1700 – 1750 range is due to the stretching vibrations of carbonyl ($>C=O$) and/or carboxyl (–COOH) species (15). Absorbed water causes the peak at about 1625 cm^{-1} (16). Absorption peaks at wave numbers below 1430 cm^{-1} are brought about by C–C and C–O stretching as well as various bending vibrations.

The region of interest is examined in Figure 2, where the effect of baking at 140 °C for periods up to 500 hr is seen. It is apparent that accelerated aging causes an increase in either carbonyl, carboxyl or both. It is also of interest to note the increased absorbance at 1625 cm^{-1} due to a higher moisture content of the film. This phenomenon of higher moisture regain has also been reported for archaeological fabrics (17).

The outcome of a one–hour water wash and of a one–hour NaBH$_4$ treatment are shown in Figure 3. It may be seen that a water wash partially decreases the absorption at the wavelength of interest. Thus, the absorbing species must be water–soluble, low–molecular–weight products that are not part of the cellulose chains. The removal of this material is also in keeping with the results of Reference 17, in which it was found that archaeological textiles showed an appreciable weight loss upon washing.

Practically all of the absorption at 1725 cm^{-1} is eliminated after NaBH$_4$ treatment, indicating that this absorption is caused by species either attached to the cellulose or too large to be removed by water. Further investigation showed that the absorbance at 1725 cm^{-1} continued to decrease with longer treatment times, reaching a very small limiting value after about three hours.

The increase in absorbance at 265 nm with baking time at 140 °C is shown in Figure 4. It has been noted previously (5) that this absorption is due to carbonyl groups. The effects of washing and treatment are shown in Figure 5. Again, it is apparent that some of the absorbance is due to water–soluble material and that the material which is not removed is affected by a reduction treatment. It was found that continued treatment reduced the absorbance in this region to about the same value as that of the control. Since NaBH$_4$ does not affect carboxy acids, the evidence from both IR and UV spectra indicates that the concentration of acid species, if there are any, is negligible in comparison to that of the carbonyl species.

The UV spectrum of the aqueous extract is also shown in Figure 5. Note that its shape is the same as that of the baked film, but that its maximum is shifted toward lower wavelengths. This shift is also seen in the washed film, but does not appear in the treated film. Further study showed that the position of the absorption peak was sensitive to aqueous treatment, and that it was generally shifted to

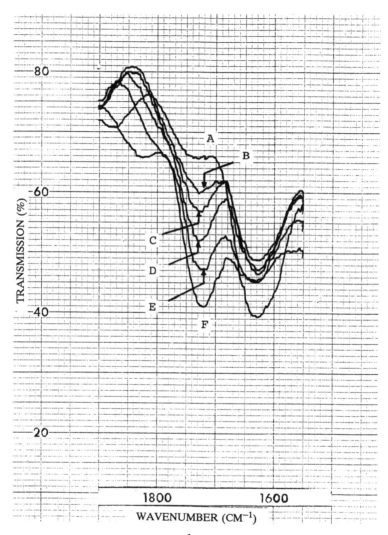

Figure 2. Decrease in 1725 cm^{-1} Percent Transmission of Cellophane Film Aged at 140 °C: (A) Unbaked, (B) 100 hr, (C) 200 hr, (D)
300 hr, (E) 400 hr, (F) 500 hr.

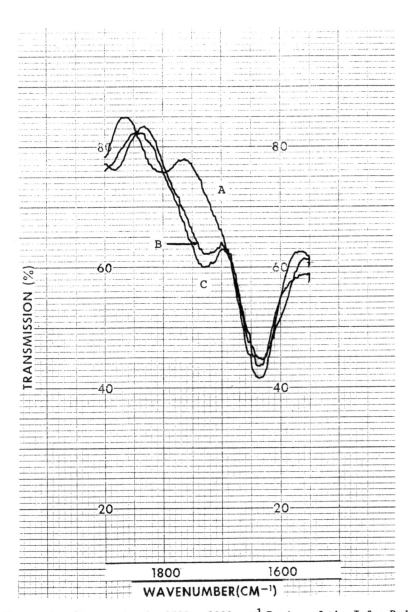

Figure 3.   Changes in the 1500 – 1900 cm^{-1} Region of the Infra-Red Spectrum of Cellophane Film (C) baked for 150 hr at 140 °C, then (B) washed or (A) treated.

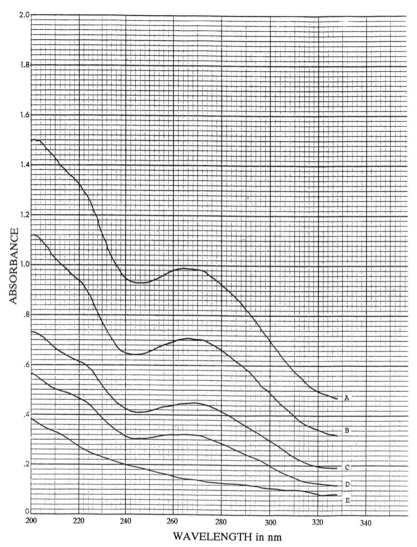

Figure 4. Increase in Ultra-Violet Absorption of Cellophane Film
Aged at 140 °C: (A) 100 hr, (B) 50 hr, (C) 30 hr, (D) 20 hr, (E)
Unbaked.

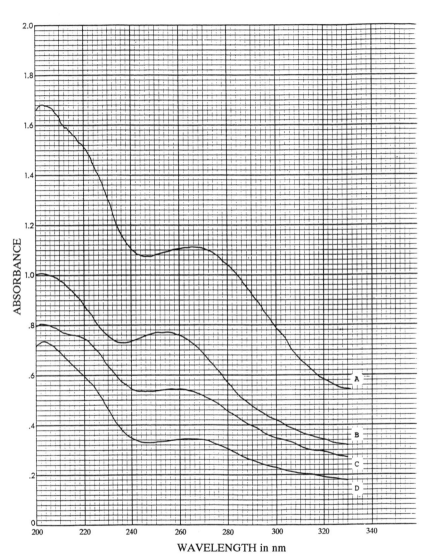

Figure 5. Changes in the Ultra-Violet Spectrum of Cellophane Film (A) baked for 150 hr at 140 °C, then (C) washed or (D) treated, and (B) water extract.

shorter wavelengths. It is not known why the absorption peak shifts, but may be due to solvation effects. Therefore, all measurements were made at the absorption maximum between 255 and 270 nm. The IR spectrum of the aqueous extract is shown in Figure 6. Note the >C=O absorption peak at about 1700 cm^{-1}. The exact composition of the extract is not known. Note, however, that (1) the extract absorbs at about 265 nm in the UV, indicating carbonyl species; (2) the extract absorbs at about 1700 cm^{-1} in the IR, indicating either carbonyl or carboxyl species; and (3) that the extract is acidic. It is likely, then, that this peak may be assigned to various low-molecular-weight degradation products that may account for the acidity of the extract. It also should be noted that this spectrum is essentially the same as that reported by Kleinert (18, 19) for the aqueous extract of naturally and artificially aged linen.

Kinetics

The change in absorbance at the maximum in the 255 – 270 nm region of films baked for various periods of time at 140, 110 and 90 °C are shown in Figure 7. All of the data were fit to curves of the form of Equation 5. The reaction rate results are summarized in Table I for both the early (up to twenty hours) and the latter portions of the curves (beyond twenty hours).

This range of temperatures, 90 – 140 °C, was chosen because it includes the boiling point of water. Since films baked at temperatures above 100 °C have a much lower moisture content than those baked at <100 °C (3, 20), it was expected that any significant changes in reaction mechanism due to the presence of water would be revealed.

Table I: Reaction Rates (x 10^{-4}) for Cellophane Film
Aged at Temperatures from 90 – 140 °C

Baking Time (hr)	Temperature (C)		
	140	110	90
0 – 20	52	9	3.0
20 – 400	20	1.5	0.4

It is apparent that the curves of Figure 7 come from the same family. As expected from the model, each of them can be fit to a first-order kinetic equation with a constant value of A_o, except for a more rapid rise in the very early portion. All of the curves must, of course, have the same value of A_o since all of the films start with the same amount of cellulose.

In addition, the shapes of the curves are the same as those found from measurements of degree of polymerization, color change and tear strength for the degradation of cotton, linen and rayon cloth artificially aged under the same conditions (12, 14, 21).

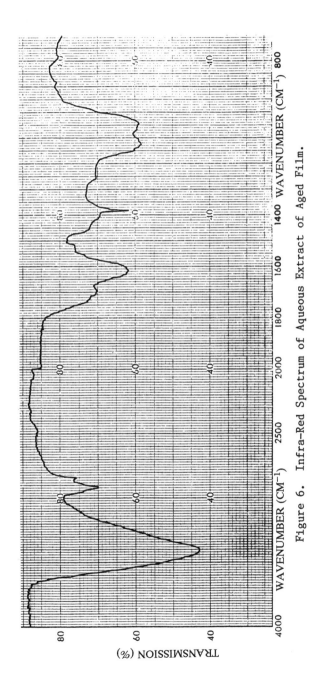

Figure 6.   Infra-Red Spectrum of Aqueous Extract of Aged Film.

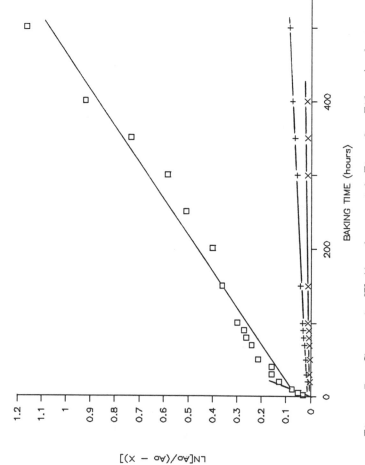

Figure 7.   Change in UV Absorbance with Time for Films Aged at
(☐) 140, (+) 110, and (X) 90 °C.

It was also determined that the shapes of the IR and UV spectra of the baked films remained the same irrespective of baking temperature; indicating that baking temperature, and thus moisture concentration, did not significantly affect the reaction mechanisms. This latter result is confirmed by work soon to be published (Mayhew, F., University of Delaware, personal communication, 1988). For long aging times, beyond twenty hours, the overall activation energy was calculated to be about 24 kcal/mole over this temperature range; in good agreement with previous work (22 - 24). For the short aging times, up to twenty hours, the overall activation energy was found to be about 17 kcal/mole.

Moisture

Since it is known that moisture increases the overall reaction rates for both paper and fabric (25 - 27), it was postulated that the early rapid rise in absorption might be due to moisture in the films. Therefore, film samples stored in air at about 55% RH for at least 24 hr prior to baking at 140 °C were compared with samples stored in a desiccator. One set of samples was taken from film as received, one set from water-washed film, and one set from film treated with $NaBH_4$. The results of this investigation are given in Table II. As before, all of the data were fit to curves of the form of Equation 5. The reaction rates are given in Table II.

Table II. Reaction Rates for Moist and Dry Films
Aged up to 20 hr at 140 °C

Film Type	Reaction Rate (x 10^{-4}/hr)	Intercept
As received, dry	46	0.023
As received, moist	52	0.020
Washed, Dry	36	0.009
Washed, Moist	37	0.011
Treated, Dry	30	0.004
Treated, Moist	31	0.005

These results indicate that for baking times up to 20 hr the rate of formation of absorbing species in the as-received films is greater than that for water-washed films, and that these exhibit a greater rate of increase in absorbance than do the treated films. At the 95% confidence level the reaction rates for each film type are significantly different. Furthermore, at 140 °C the differences in rates between films stored in a desiccator and those stored in moist air are not statistically significant.

The films that were not treated with sodium borohydride show an intercept on the vertical axis that is statistically different from zero; and that offset is greater for the as-received films than for the water-washed films. This latter result indicates that the films contain water-soluble, low-molecular-weight materials that can be, at least partially, removed as well as carbonyl species that can be reduced by the $NaBH_4$ treatment.

The results of similar experiments performed at the lower temperatures, 110 and 90 °C, were not precise enough to determine if the differences between rates were statistically significant.

It should be noted that, since the as-received, water-washed and borohydride-treated films exhibit the same IR and UV spectra upon baking, the products formed upon accelerated aging at 140 °C appear to be essentially the same as those found in the naturally oxidized films.

It may be concluded that the rapid reaction rate in the first twenty hours of aging is due, not to moisture, but to oxidized material present in the as-received film.

Relation to Natural Fibers

That the changes in IR and UV absorbance are related to chemical changes in the film is exhibited in Figure 8, where the color change is shown as a function of baking time at 140 °C (Mayhew, F., University of Delaware, personal communication, 1988). It is readily apparent that color change in the films follows the same kinetics as changes in UV and IR absorbance, although at different rates.

In addition, it was earlier shown (28) that changes in color and tear strength of cotton and rayon cloth could be modeled in the same manner. The kinetics of color change of the cloths exhibited the same behavior as the Cellophane films in this study. That study, conducted at 150 °C, gave a pseudo reaction rate for color change of 0.03 units/hr. Using the value for activation energy found in this work of 24 kcal/mole, yields a prediction of 0.021 units/hr for Mayhew's data set. The actual value is 0.015 which, considering the precision of activation energies, is remarkably good.

Pre-Aging

The data shown in Table II indicate that films treated with sodium borohydride aged more slowly than water-washed films. In order to ascertain if this outcome could be achieved with old as well as with new materials, films were artificially aged by baking for 20, 40, 80 or 150 hr prior to either a one-hour water soak or a one-hour treatment. The results of this study are shown in Table III.

The difference between rates was confirmed by a t-test at the 95% level. It is readily apparent that for films aged up to 80 hr, the treatment is superior to the wash. For very old films, however, it should be noted that the $NaBH_4$ treatment is not as efficacious, indic-

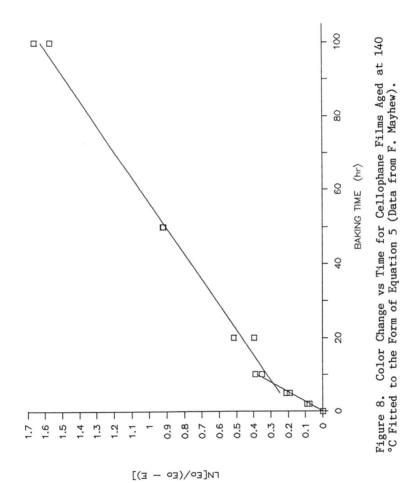

Figure 8. Color Change vs Time for Cellophane Films Aged at 140 °C Fitted to the Form of Equation 5 (Data from F. Mayhew).

ating that the extent of oxidation is such that a one-hour treatment is not long enough to alleviate the effects of aging. It was found that extended treatment of films baked for up to 500 hours would further decrease the absorbance.

Table III. Reaction Rates (x 10^{-4}) for Pre-Aged Washed and Treated Films Baked at 140 °C for up to 450 hr

Pre-Aging Time (hr)	Reaction Rate	
	Washed	Treated
0	19.9	
20	18.0 *	13.4
40	19.5 *	16.6
80	15.4 *	12.2
150	14.3	15.7

* Significantly different at 0.05 level

CONCLUSIONS

It has been shown that changes in the UV and IR absorbance of unplasticized Cellophane films subjected to accelerated aging in a dry oven at 140 °C follow the behavior predicted by a first-order kinetic model, except for deviations in the early aging period, and that these deviations are most likely caused by oxidation products in the films.   It has also been shown that, for Cellophane films, the changes in UV and IR absorbance follow the same kinetics as color change, and that these kinetics are nearly identical with those for rayon and cotton cloths aged under similar conditions.
     It was further shown that treatment with sodium borohydride (1) effectively reduces the extent and rate of degradation of both new and aged Cellophane films, in agreement with previous reports on cotton cloth, and (2) reduces both UV and IR absorbance to negligible levels; indicating that acid moieties on the cellulose chains, if present, are in negligible quantities.
     In addition, it was noted that the IR spectra of aged Cellophane films and the aqueous extract of the aged films are essentially the same as those of naturally aged linen.   Therefore, this study provides strong evidence that (1) unplasticized Cellophane films may be used as a model material for studies of the aging of cellulosic fibers, (2) measurement of the absorbance of these films at about 265 nm is a satisfactory method for following the kinetics of accelerated aging, (3) at least for Cellophane film, oven aging may be conducted at temperatures between 90 and 140 °C without change in reaction mechanisms, and (4) the darkening and strength loss exhibited by cellulosic textiles is more likely affected by carbonyl rather than by carboxyl species.

Acknowledgment

The authors thank Professor Frances Mayhew for supplying the data for Figure 8.

Literature Cited

1. Peters, R. H. Textile Chemistry, Volume 1; Elsevier Publishing Co.: New York, 1963; p 179.
2. Major, W. D. Tappi 1958, 41, 530 - 37.
3. Nikitin, N. I., The Chemistry of Cellulose and Wood; Trans. by J. Schmorak; Israel Program for Scientific Translations: Jerusalem, 1966; Chapter VIII.
4. Arney, J. S.; Jacobs, A. J. Tappi 1979, 62, 89 - 91.
5. Lewin, M.; Epstein, J. A. Journal of Polymer Science 1962, 58, 1028 - 37.
6. Nevell, T. P. In Cellulose Chemistry and Its Applications; Nevell, T. P.; Zeronian, S. H., Eds.; Wiley and Sons: New York, 1985; Chapter 6.
7. McBurney, L. F. In Cellulose and Cellulose Derivatives, Part I; Ott, E.; Spurlin, H. M.; Grafflin, M. W., Eds.; Interscience, New York (1954).
8. Sharples, A. In Cellulose and Cellulose Derivatives; Bikales, N. M.; Segal, L., Eds.; Wiley-Interscience: New York, 1971; Part V, p 994.
9. Nevell, T. P. In Cellulose Chemistry and Its Applications; Nevell, T. P.; Zeronian, S. H., Eds.; Wiley and Sons: New York, 1985, Chapter 11.
10. Tang, L. In Historic Textiles and Paper Materials: Conservation and Characterization; Needles, H. L.; Zeronian, S. H., Eds.; American Chemical Society: Washington, D. C., 1986, p 427 -41.
11. Browning, B. L.; Wink, W. A. TAPPI 1968, 51, 156 -63.
12. Block, I.; Kim, H. K. In Historic Textiles and Paper Materials: Conservation and Characterization; Needles, H. L.; Zeronian, S. H., Eds.; American Chemical Society: Washington, D. C., 1986, p 411 -27.
13. Cardamone, J. M.; Brown, P. In Historic Textiles and Paper Materials: Conservation and Characterization; Needles, H. L.; Zeronian, S. H., Eds.; American Chemical Society: Washington, D. C., 1986, p 41 -76.
14. Feller, R. L.; Lee, S. B.; Bogaard, J. In Historic Textiles and Paper Materials: Conservation and Characterization; Needles, H. L.; Zeronian, S. H., Eds.; American Chemical Society: Washington, D. C., 1986, p 329 - 347.
15. Zhbankov, R. G. Infrared Spectra of Cellulose and Its Derivatives; Consultants Bureau: New York, 1966.
16. O'Connor, R. T. In Analytical Methods for a Textile Laboratory, Second Edition; Weaver, J. W., Ed., American

Association of Textile Chemists and Colorists: Research Triangle Park, 1968.

17. Hersh, S. P.; Hutchins, J. K.; Kerr, N.; Tucker, P. A. In Conservazione e Restauro dei Tessili: Proceedings of the International Conference, Como 1980; Centro Italiano per lo Studio della Storia del Tessuto (CISST) – Lombardy Section; Milan, 1980; p 87 – 95.

18. Kleinert, T. N.; Marraccini, L. M. Svensk Papperstidning 1966, 69, 159 – 60.

19. Kleinert, T. N. Holzforschung 1972, 26, 46 – 51.

20. Urquhart, A. R.; Williams, A. M. J. Text. Inst. 1924, 15, T433 – 42 and T559 – 72.

21. Block, I. J. Am. Inst. Conserv. 1982, 22, 25 – 36.

22. Kim, H. K. M. S. Thesis, University of Maryland at College Park, 1985.

23. Roberson, D. D. TAPPI 1976, 59, 63 – 69.

24. Block, I. Preprints of the 7th Triennial Meeting of the ICOM Committee for Conservation, 1984, p 84.9.7 – 84.9.10, (1984).

25. Richter, G. A.; Wells, F. L. TAPPI 1956, 39 603 –08.

26. Browning, B. L.; Wink, W. A. TAPPI 1968, 51, 156 – 63.

27. Graminski, E. L.; Parks, E. J.; Toth, E. E. In Durability of Macromolecular Materials; Eby, R. K., Ed; ACS Symposium Series No. 95; American Chemical Society: Washington, DC, 1979; pp 341 –69.

28. Block, I. J. Am. Inst. Conserv. 1983, 22(1), 30 –36.

RECEIVED March 6, 1989

Chapter 15

Nondestructive Evaluation of Aging in Cotton Textiles by Fourier Transform Reflection–Absorption Infrared Spectroscopy

Jeanette M. Cardamone

Department of Clothing and Textiles, College of Human Resources, Virginia Polytechnic Institute and State University, Blacksburg, VA 24061

Specular Reflectance FTIR has been used to follow the chemical changes in cellulose when cotton cloth was aged for 2, 4, 7, 11, 14, 23, and 31 hours at $190^{\circ}C$ in air and in nitrogen. The infrared absorptions of carbonyl and carboxylate groups were measured from the 1600–1750 cm-1 region. After 31 hours of aging in air and in nitrogen, the extent of carboxylate formation was approximately the same although oxidation proceeded more rapidly in air. The rate of aging in nitrogen proceeded at a uniform rate over 31 hours whereas in air, a leveling off point was reached after 11 hours. The dependence of the formation of carboxylate on the accessibility of OH groups is discussed. These results indicate the exceptional use of Specular Reflectance absorbance infrared spectroscopy for rapid, sensitive, and nondestructive detection of oxidation degradation in cotton textiles.

The aging of cellulosic textiles includes a complex set of reactions involving oxidation, hydrolysis, crosslinking, and chain scission. (1–3) Slow degradation at room temperature is cumulative and can be followed by a chemical change in the cellulose structure or by a physical change in some measured parameter. Often artificial heat-aging is used to simulate the natural aging process assuming that similar changes occur. (4–7) These changes from thermal aging in air, in nitrogen, or in vacuo affect cellulose structural morphology, chemical accessibility, and the extent of degradation. (8–11)

General Infrared Spectroscopy Background

Infrared spectroscopy is recognized as a universal instrumental method for analyzing changes in chemical structure. (12–13) There are a variety of infrared techniques which include Transmission,

0097–6156/89/0410–0239$06.00/0

Diffuse Reflectance, Attenuated Total Reflection or Multiple
Internal Reflection, Photoacoustic (PAS), Photothermal Beam
Deflection, Specular Reflection Absorption, and forensic
applications with the diamond cell and the Fourier transform
infrared (FTIR) microscope. In museum laboratories, FTIR
applications have been used for problems of identification and
degradation in art and archeology. (14)

Several spectroscopic approaches have been used to examine
textiles. In Transmittance FTIR an infrared transparent sample may
be suitable for placement directly in the infrared beam or may be
ground or pulverized, mixed with KBr or KCl and pressed as a pellet
or disc. The sample can be mulled or deposited as a film. (15-17)
The Diffuse Reflectance technique requires dispersing the finely
divided substrate in a scattering salt such as KBr or KCl. This
method has been applied to examine highly scattering glass fibers.
(18) Attenuated and Multiple Reflectance FTIR have been used to
study finish treatments on textiles. (19-20) These reflectance
methods are complicated because yarn interlacings and cloth
intersticies prevent close contact with the reflecting prism
surfaces. Other FTIR applications include Photothermal Beam
Detection (21) and the use of diamond cells for microsampling of
micrograms of museum objects. (22-23)

Photoacoustic FTIR spectroscopy is ideal for measuring
absorptions from solid samples because it is unaffected by
scattering radiation. (24) The theory and technique are described
elsewhere. (25-26) In Figure 1, wavenumber versus relative
photoacoustic intensity for cotton print cloth (#400 U Testfabrics,
Inc.) in the bottom spectrum is compared to the spectrum of
naturally aged cotton cloth from 1250 - 1300 A.D. (26). The spectral
changes with aging shown in the 1600 - 1750 cm-1 region are typical
of those found by others. (27-28) Cut samples were analyzed from the
cotton control. Yarn fragments, 0.5 cm long were analyzed from the
naturally aged cotton fabric.

Specular reflectance FTIR has been used to examine solids and
semi-solids, adhesives, and coatings. There are three systems based
on reflectance angle orientations: grazing for thin films, surface
contaminates, and lubricant films; variable angle for thin films and
highly reflective surfaces and coatings, and fixed angle for
coatings and reflective surfaces. Surface applications include
following photochemical changes of bulk and polymer coated metal and
metal oxide interfaces (29), oxidation degradation of the surface
layer of a composite sample (30), and evaluation of degradation and
reactions at metal polymer interfaces due to environmental factors.
(31) Reflectance FTIR has been used as a nondestructive and
reproducible method for determining the amount of lubricant on the
surface of magnetic discs. (32) A search of the textile literature
showed no application of specular reflectance FTIR for textiles.

The Specular Reflectance FTIR study presented here follows from
earlier work where methodology using property-kinetics was developed
to estimate extent of degradation. (33) Specular reflectance FTIR
was examined for possible use in applying the methodology
nondestructively by monitoring carboxylate development for future
correlation with strength loss. Although the Photoacoustic study of
naturally and artificially heat-aged cottons proved to be a rapid

and sensitive method for monitoring cellulose degradation; the
specular reflectance method was investigated because it requires no
sample taking.

Experimental

Cotton print cloth was heat-aged at 2 ,4, 7, 11, 14, 23, and 31
hours in air and in nitrogen at 190°C. The procedure is described
elsewhere.(33) Infrared absorption was measured as a function of
wavelength to distinguish carbonyl and carboxylate groups which
absorb from 1600 - 1750 cm-1. The Fourier transform infrared
spectrometer is described in detail with theory and methods in
Griffith and de Haseth's book on the subject. (34) The Specular
Reflectance spectra presented here show high signal-to-noise and
were collected by placing a 1 3/4" x 2" cloth sample on top of the
1/2" diameter opening of a reflectance sample cell holder which
was placed in the infrared beam. The sample cell was configured with
two fixed mirror planes meeting at an apex below the opening which
directed the infrared beam horizontally and upward at an incidence
angle, being directed to the detector by the second mirror plane.
All measurements were of absorbance by peak height values. A Mattson
Sirius model 100 FTIR interferometer and computerized data
collection, storage, and data manipulation system was used to
collect all spectra into individual data files. A mercury cadmium
telluride detector with transmission window of 4000 to 500 cm-1 was
used. A computer program was written to co-add 16 mirror scans in
the foward direction and 16 in reverse. Each single pass of the
mirror recorded 256 scans for a total of 8,192 data collections each
at 2 cm-1 resolution. These data collections were collected through
a 16 bit A/D converter and were signal averaged into 32 bit
words for further computations.

In Figure 2A a mirror spectrum was collected to represent the
background. A mirror spectrum was collected at the beginning of
each session of data collection and were were used to ratio
singlebeam spectra. This normalization corrected for instrument
variability. Figure 2B is the Fourier transform of the interferogram
(singlebeam spectrum) of the unheated control sample. Figure 2C is
the absorbance spectrum obtained by ratioing B to A. All other
absorbance spectra were obtained by ratioing a singlebeam spectrum
to the mirror spectrum. These other spectra are represented by
Figure 3A, 2 hours in air, 3B, 31 hours in air, 4A, 14 hours in air,
and 4B, 14 hours in nitrogen. In Figure 5 the development of the
carboxylate band at 1730 cm-1 is ratioed against the CH stretching
band in the 2900 cm-1 region. The 2900 cm-1 band was used as a
second normalization standard for each sample since it showed no
appreciable change with aging. Normalizing to this internal
standard provided a correction for changes in reflectivity due to
changes in sample positioning. In Figure 6 the 1730 cm-1
carboxylate band is compared to the OH stretching absorptions which
represent intermolecular hydrogen bonding at 3300 cm-1 for an
indication of the interdependence of these functional groups during
aging.

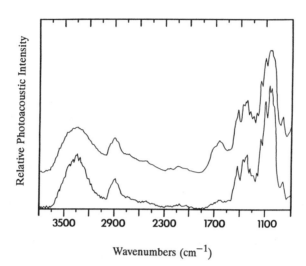

Wavenumbers (cm^{-1})

Figure 1 - Photoacoustic FTIR spectra of fibers from new
cotton cloth (lower Spectrum) and approximately 700 year old
cotton (upper spectrum). The aged cotton fiber was sample
#3019 (circa A.D. 1250-1300). Both spectra were obtained by
averaging 1000 interferometer scans. Note the presence of new
absorption bands in the C=O region in the aged sample.
* Reproduced with permission from reference 26. Copyright 1987
Textile Research Journal.

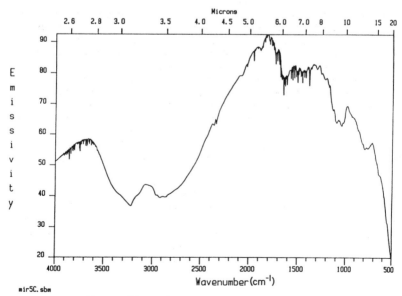

ir5C.sbm

Figure 2A - Mirror Background Spectrum

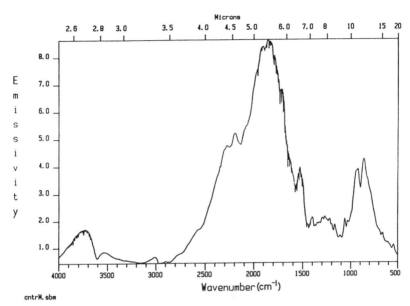

cntrM. sbm

Figure 2B - Singlebeam Spectrum of Unheated Cotton Cloth

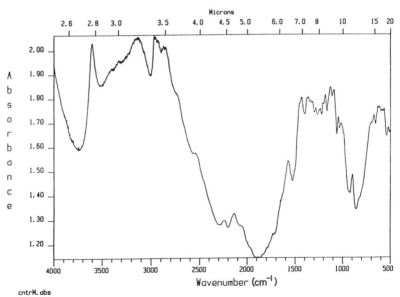

cntrM. abs

Figure 2C - Absorbance Spectrum of Unheated Cotton Cloth

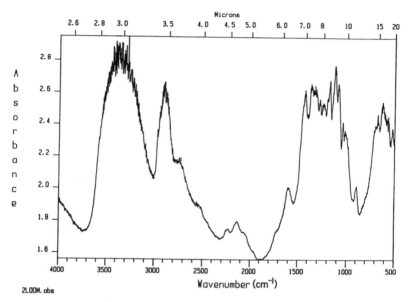

2LOOM. obs

Figure 3A – Cotton Cloth Aged for 2 hours in air at 190°C

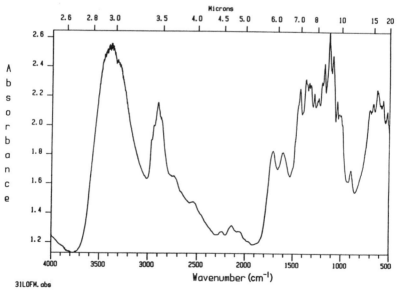

31LOFM. obs

Figure 3B – Cotton Cloth Aged for 31 hours in air at 190°C

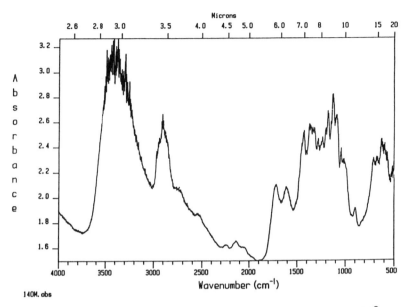

Figure 4A – Cotton Cloth Aged for 14 hours in air at 190°C

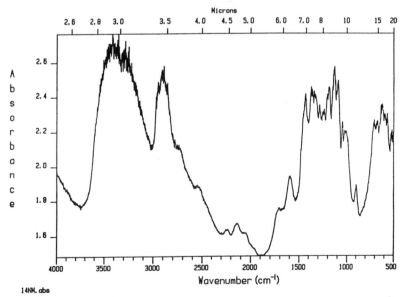

Figure 4B – Cotton Cloth Aged for 14 hours in nitrogen at 190°C

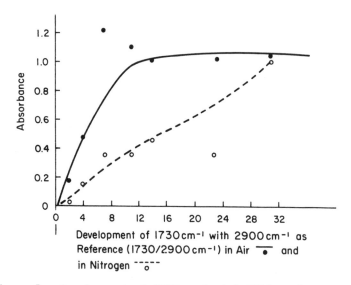

Figure 5 – Development of 1730 cm–1 with 2900 cm–1 as
         Reference (1730/2900 cm–1) in Air and in Nitrogen

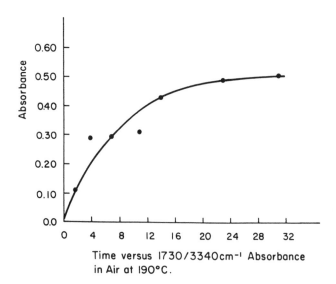

Figure 6 – Time versus 1730/3340 cm–1 Absorbance in Air at 190°C

Discussion and Results

There is a wide variety of complex cellulose oxidation reactions. Oxidation products, oxycelluloses, can form from selective oxidations of the C-6 OH, C-2 OH, and C-3 OH groups to carbonyl and carboxyl groups (35). These OH groups on cellulose have different reactivities. It can be reasoned that oxidation will cause reducing end group conversion from OH at the reducing end of a polymer chain to aldehyde to acid carboxylate (36). Different reactivities and accessibilities can be found in different parts of the fiber. An extensive body of literature has been amassed to describe the effects of various oxidizing agents including 21% oxygen in air on the formation of oxycelluloses under widely diverse conditions of time, temperature, and concentration.

During low temperature degradation of cellulose the glycosyl units, C(1) -O -C(4'), decompose with the formation of water, carbon monoxide, and carbon dioxide. (37) With the cleavage of the gylcosyl units, shorter cellulose chains form which can lead to a loss in tensile strength. Certain measured parameters such as weight loss, degree of polymerization, strength retained, and color change have been used to follow the kinetic effects of heat-aging in air. (8,38) The thermal degradation of cellulose is thought to fit an overall pattern or series of steps which describe the rate of attack on the glycosyl linkages. The first initial stage affects weak linkages. This is followed by a slower rate where these linkages in the amorphous regions are affected. In the final stage the slowest rate of attack is proposed for the crystalline regions where a leveling off degree of polymerization (LODP) is described. (39)

Specular Reflectance FTIR Spectra.   When the spectra are examined for the effects of aging, the unheated cotton control in Figure 2C contains the sharp band at 3600 cm-1 assigned for free OH stretching (non-hydrogen- bonded OH).(40) The sample was free of water and the spectrum was collected after a long purge of the optical bench where the samples were stored under the drying effect of liquid nitrogen used to cool the detector. All samples were handled inside a glove bag sealed to the opening to the sample compartment.

By contrast, the broad hydrogen bonded absorption at 3300 cm-1 in Figure 3A indicates that after two hours of aging there are definite morphology changes brought about by inter-and intra-molecular hydrogen bonding which can lead to crosslinking.

Another critical spectral region is 1630 cm-1 where water absorption bands have been reported (41) In this study this band develops from chemical change and not from absorbed water. This is supported by the spectral changes in 3360 and 1630 cm-1 which do not procede on the same time frame, consequently the chemical changes in these groups are not due to the same chemical species. The assignment of the 1630 cm-1 absorption as carbonyl is consistant with the spectra of -glucose in its aldohexose form. (41)

When the Specular Reflectance spectra in Figures 3A and 3B are compared to the PAS spectrum in Figure 1, the Reflectance spectra show somewhat better spectral detail because of a higher resolution (2 cm-1 versus 4 cm-1, and a higher number of data collections, 8,192 versus 1000).

    Compared to the PAS spectra of cotton cloth, in the Specular
Reflectance spectra, the signal-to-noise is higher and any noise is
low enough to show clearly the spectral features described here.
Even though there are differences due to the change in sampling
conditions, the spectra can be compared for a one-to-one comparison
and identification of absorption bands. Note however that in the
PAS spectrum relative intensities are displayed and that the
Specular   Reflectance   spectra   show   an   absolute   response   and
measurement which leads to a better possibility for quantification
and a more rigorous treatment of the data.
    The following summarizes the changes revealed by Specular
Reflectance spectroscopy when cotton cloth is heat-aged:
.. Air aging after 31 hours leaves the cloth in a friable and
    brittle state where the extent of degradation is very high.
.. In nitrogen after 31 hours, carboxylate forms to the same extent
    as after 31 hours in air (Figure 5).
.. There is a difference in the rate of aging when the air and
    nitrogen environments are compared. There is greater carboxylate
    absorption in air at 14 hours than in nitrogen in 14 hours.
.. After 11 hours of aging in air, a leveling off point for
    carboxylate development is reached.
.. Up to 31 hours of aging, there is no leveling off for
    carboxylate development when cotton cloth is aged in nitrogen.
.. When the 1730/3340 cm-1 absorption ratios are plotted versus
    wavenumber the rate of carboxylate formation is faster than the
    rate of change in OH (Figure 6). In fact there is little change
    in the OH region over 2 to 31 hours of aging.

Conclusions

The heat-aging of cotton cellulose in nitrogen involves carbonyl and
carboxylate group formations which can be monitored by the infrared
absorptions at 1600 - 1750 cm-1 with Specular Reflectance FTIR. Not
all principal bands undergo change. For example, there is little
change in the regions for OH absorption at 3340 cm-1, CH at 2900
cm-1, and carbonyl at 1630 cm-1 over 2 to 31 hours of aging in air
and in nitrogen. The pronounced increase in the 1730 cm-1 band
for carboxylate indicates that the oxidation may not form by direct
conversion of these other functional groups whose absorptions change
only slightly. It is possible, however, that there may be a steady
supply of these groups which becomes available if these groups are
involved in this chemical change. For example, the accessibility of
the OH groups in the amorphous regions may increase with aging. By
Figure 6, more OH groups could become available at a slower rate
than the formation of the carboxylate groups. It can be reasoned
that a possible source for the increase in 1730 cm-1 absorption is
the conversion of the reducing aldehydic end groups to acid groups.
Other sources of carboxylates may be by glycosyl bond cleavage and
the subsequent formation of shorter chains with end groups having
OH and aldehyde which oxidize to carboxylate.
    The leveling off point in Figure 6 has been found by others and
has been characterized at the physical level as the limit of
acceptable strength. (39) Leveling off has been explained as the end

of the attack in the amorphous regions and the resistivity to
chemical attack of the remaining highly ordered microcrystalline
regions (39).
The spectral data collected for Figure 5 support what others
have found, that air-aging procedes faster than nitrogen-aging but
that in both environments, an endpoint can be reached. (8) The
oxycelluloses which form in cellulose with aging are therefore time
and temperature dependent. The rate at which they develop can be
controlled to a limited degree by choosing a nitrogen environment
to slow the process. With this exceptional use of Specular
Reflectance FTIR, it may be possible to follow the aging process
nondestructively and to estimate the extent of degradation which
corresponds to the end of a textile's useful life.

Acknowledgments

Professor James O. Alben, Director of the Biospectroscopy
Laboratory, The Ohio State University (NIH Grants: HL28144 and
RR01739) provided the facility and training for the author. Salary
and research support were provided in part by state and federal
funds appropriated to the Ohio Agricultural Research and Development
Center and to the Virginia Agricultural Experiment Station.

Literature Cited

1. Hebeish, A.; El-Aref, A. T.; El-Alfi, E. A.; and El-Rafie, M.
   H. J. Appl. Polm. Sci. 1979, 23, 453-462.
2. Philipp, B.; Baudisch, J.; and Ruscher, C. TAPPI J. 1969, 52,
   (4), 693-698.
3. McGinnis, G. D.; Shafizadeh,; Casey, J. P. Ed. In Pulp and
   Paper Chemistry and Chemical Technology, 3rd ed.
   Wiley-Interscience: New York, 1980; Vol 1, Chapter 1.
4. Rasch, R. H.; Scribner, B. W. Res. Nat. Bur. Stand., 1933, 11,
   (5), 727-732.
5. Scribner, B. W. J. of Res. Nat. Bur. Stand., 1939, 23 (3),
   405-413.
6. Richter, G. A.; Wells, F. L. TAPPI J, 1956, 39, 603- 608.
7. Block, I. J. Amer. Inst. Conserv., 1982, 22, 25-36.
8. Shafizadeh, F.; Bradbury, A. G. W. J. Appl. Polym. Sci., 1979,
   23, 1431-1442.
9. Parks, E. J. TAPPI J., 1971, 54 (4), 537 - 544.
10. Fairbridge, C.; Ross, R. A.; Sood, S. P. J. Appl. Polym. Sci.,
    1978, 22, 497 - 510.
11. Zeronian, S.H. In Cellulose Chemistry and Technology; ACS
    Symposium Series 48; American Chemical Society: Washington,
    D.C. 1977; pp. 189-205.
12. Morris, N.M.; Berni, R.J. In Polymers for Fibers and
    Elastomers; Arthur, J. C., Jr., Ed.; ACS Symposium Serives
    No. 260, American Chemical Society: Washington, D.C., 1984;
    62-74.
13. Yang, C.Q.; Bresee, R.R.; Fately, W.G.; Perenich, T. A. In The
    Structures of Cellulose; Atalla, R. H. Ed.; ACS Symposium
    Series No. 340, American Chemical Society: Washington, D.C.,
    1987; 214- 232.

14. Van Zeist, L; von Endt, D.W.; Baker, M. T. Proc. 2nd International Conference on Non-Destructive Testing, Microanalytical Methods and Environment Evaluation for Study and Conservation of Works of Art, Perugia, 1988; I/30.1 - I/30.17.

15. Geary, M. A.; Hawkins, T. W. Text. Chem. Col., 1981, 13, 135-237.

16. Abrahams, H. H.; Edelsetin, S. M. Am. Dyest. Rep. 1964. 53, 19-25.

17. Law, M. J. D.; Baer, N. S. Stud. Conser., 1977, 22. 116-128.

18. Maulhardt, H.; Kunath, D. A. Appl. Spectr., 1980, 34, (3), 383-385.

19. American Association of Textile Chemist and Colorist. Text. Chem. Col., 1973, 15, 30-35.

20. Flor, R. V.; Prager, M. S. J. Text. Inst. 1982, 73, 138-141.

21. Low, M. J. D.; Verlashkin, P. G. Appl. Spectr. 1986, 31, 77-82.

22. McCawley, J. C. Proc. ICOM Comm. Conserv. 4th Trennial Meeting, 1975, 75/45. 1-12.

23. Laver, M. E.; Williams, R. S. J. Int. Inst. Conserv. - CG, 1978, 3,(2), 34- 39.

24. Freeman, J.; Friedman, R. M. J. Phys. Chem., 1980. 84, 315-319.

25. Rosencwaig, A. Anal. Chem., 1975. 47, 592A -604A.

26. Cardamone, J.M.; Gould, J. M.; Gordon, S. H. Text. Res. J. 1987, 57, 235 - 239.

27. Higgins, H. G. J. Polym. Sci., 1958, 28 (118), 645 - 648.

28. Kleinert, T. N. Holzforschung, 1972, 2, 46 - 51.

29. Webb, J. D.; Schissel, P.; Thomas, T. M.; Pitts, J. R.; Czandra, A. W. Solar Energy Materials, 1984, 11, 163 -175.

30. Takase, A.; Yashida, H.; Yamamoto, T. J. Mat. Sci. Lett., 1985. 4, 982 - 984.

31. Sergides, C. A.; Chughtai, A. R.; Smith, D. M. Appl. Spectr. 1985, 39, 735 - 737.

32. Walder, F.T.; Virdine, D. W.; Hansen, G. C. Appl. Spectr., 1984, 38, 782 - 786.

33. Cardamone, J. M.; Brown, P. In History Textile and Paper Materials Conservation and Characterization; Needles, H. L.; Zeronian, H. L. Eds.; Advances In Chemistry Series No. 212. American Chemical Society: Washington, D.C., 1986, 41 - 75.

34. Griffiths, P. R.; de Haseth, J. A. Fourier Transform Infrared Spectroscopy, Wiley; New York, 1986.

35. Hartsuch, B. E. Introduction to Textile Chemistry. Wiley; New York, 1950, 139.

36. Peters, R. H. Textile Chemistry, Vol 1. The Chemistry of Fibers, Elsevier; New York, 1963, 183.

37. Shafizadeh, F. Thermal Degradation of Cellulose; In Cellulose Chemistry and Its Application, Nevell, T. P.; Zeronian Eds.; Wiley; New York, 1987, 267.

38. Block, I.; Kim, H. K. In Historic Textile and Paper Materials Conservation and Characterization; Needles, H. L.; Zeronian, S. H. Eds.; Advances In Chemistry Series No. 212; American Chemical Society: Washington, D.C., 1986, 412-425.

39. Feller, R. L.; Bogaard, J. In <u>Historic Textile and Paper Materials Conservation and Characterization</u>; Needles, H. L. J.; Zeronian, S. H. Eds. Advances In Chemistry Series No. 212; American Chemical Society: Washington, D.C., 1986, 330 - 347.
40. Berni, R. J. and Morris, N. M. <u>Analytical Methods for a Textile Laboratory</u>, Weaver, J. W. Ed.; American Association of Textile Chemists and Colorists, 1984, 266.
41. Zhbankov, R. G. <u>Infrared Spectra of Cellulose and Its Derivatives</u>, Consultants Bureau, New York, 1966, 41.

RECEIVED January 18, 1989

INDEXES

Author Index

Ballard, M., 134
Becker, M. A., 94
Berndt, Harald, 81
Blair, C., 134
Block, Ira, 220
Bogaard, J., 54
Butler, C. E., 34
Cardamone, Jeanette M., 239
Clements, D. W. G., 34
DeSimone, Laura, 220
Feller, R. L., 54
Ginell, William S., 108
Green, Sara Wolf, 168
Hansen, Eric F., 108
Hengemihle, Frank H., 63
Hersh, S. P., 94
Hon, David N.-S., 13

Indictor, N., 134
Jennings, T., 143
Kerr, N., 143
Koestler, R. J., 134
Lee, S. B., 54
Méthé, E., 143
Millington, C. A., 34
Needles, Howard L., 159
Nowak, Kimberly Claudia J., 159
Priest, D. J., 2
Santamaria, C., 134
Schweppe, Helmut, 188
Shahani, Chandru J., 63
Tucker, P. A., 94
Weberg, Norman, 63
Zeronian, S. Haig, vii

Affiliation Index

BASF Aktiengesellschaft, 188
Carnegie–Mellon University, 54
Clemson University, 13
Getty Conservation Institute, 108
Library of Congress, 63
Metropolitan Museum of Art, 134
North Carolina State University, 94
Textile Museum, 168
The British Library, 34

University of Alberta, 143
University of California—Berkeley, 81
University of California—Davis, 159
University of Manchester Institute
    of Science and Technology, 2
University of Maryland, 220
University of Surrey, 53
Virginia Polytechnic Institute
    and State University, 239

Subject Index

A

Abrasion resistance, effect of deacidifying
    agents, 151,152f,153,154f
Absorption coefficient
    description, 82
    physical interpretation, 84
Accelerated aging of paper, effect of
    variations in relative humidity, 63–78
Acid degradation, effective means of book
    protection, 23–24
Acid formation in paper
    aging characteristics, 16
    chief source, 16

Acidity of book papers, cause of
    deterioration, 13
Acidity of paper, definition, 3
Adhesives, tapa cloth treatment, 175
Adlerdalmatika
    identification of madder dyes, 208,210f
    photograph, 206,207f
    TLC, 206,208
Alkaline-deacidifying-agent effect on
    naturally aged cellulosic textiles
    abrasion resistance of
        fabrics, 151,152t,153,154f
    accelerated aging procedure, 146
    aqueous-extract pH of
        fabrics, 149,150t,151

Alkaline-deacidifying-agent effect on
 naturally aged cellulosic textiles—*Continued*
 color changes for accelerated aged
   fabrics, 147,149*t*
 color changes for new and naturally aged
   fabrics, 147,148*t*
 color changes on dyed fabrics, 149,150*t*
 deacidifying-agent preparation, 145–146
 description of undyed textiles, 144,145*t*
 effect of mild abrasion on cotton
   fibers, 153,155*f*
 fabric preparation, 144–145
 moisture regain of fabrics, 151,152*t*
 scanning electron microscopy,
   153,154–155*f*,156
 statistical analysis, 147
 stiffness of fabrics, 151,152*t*
 test methods, 146–147
Alkyl ketene dimers, hydrolysis
 reactions, 6,7*f*
Alkyl succinic anhydrides
 hydrolysis reactions, 6,7*f*
 sizing reactions, 6,7*f*,8
Alum, source of acidity in paper, 16–17
Aluminum sulfate method of internal sizing
 description, 2
 problem with acid hydrolysis, 3
Alum–rosin sizing, advantages and
 disadvantages, 15
Amino nitrogen content, measurement of
 degradation, 99
Ammonia content, measurement of
 degradation, 99–100
Aqueous deacidification
 advantages and disadvantages, 18
 description, 18
Artificially aged cellophane film
 accelerated aging process, 221–222
 borohydride treatment, 222
 color change vs. time, 232,233*f*
 effect of baking time on IR
   spectra, 225,226*f*
 effect of baking time on UV
   spectra, 225,227*f*
 effect of moisture, 231*t*,232
 effect of washing and treatment on UV
   spectrum, 225,227*f*,228
 effect of water wash, 225,226*f*
 IR spectra, 223,224*f*,225
 IR spectrum of aqueous extract, 228,229*f*
 kinetics, 222–223,228–231
 procedure for extracts, 222
 reaction rates, 228*t*
 reaction rates for moist and dry
   films, 231*t*
 relationship to natural fibers, 232,233*f*
 sodium borohydride vs. water
   washing, 232,234*t*

Artificially aged cellophane film—*Continued*
 spectrophotometry, 222
 UV absorbance vs. time at different
   temperatures, 228,230*f*,231
Attenuated-reflectance Fourier-transform IR
 spectroscopy, evaluation of textiles, 238

B

Backing and tear repair, tapa cloth
 treatment, 174–175
Bleached pulp
 chain breaking during extensive continuous
   exposure, 59,61*f*,62
 formation of hot-alkali-soluble matter
   during daylight exposure, 56,57*f*
 percentage of links broken during daylight
   exposure, 56,58*f*,59
 yellowing rates, 87–88,89*f*
Book deacidification
 aqueous deacidification, 18
 classifications, 18
 development, 17
 gas deacidification, 20
 nonaqueous deacidification, 19–20
Book preservation
 deacidification processes, 13–30
 integrated program, 28–29
Breaking strength, measurement of
 degradation, 99
Broke paper, definition, 10

C

Calcium carbonate fillers, applications, 5–6
Cellophane film
 accelerated aging, 221–222
 preparation, 221
 similarity to cellulose, 221
Cellulose
 composition, 220
 deterioration, 220
 low-temperature-degradation mechanism, 245
Cellulose degradation
 kinetics, 222–223
 mechanism, 221
Cellulose oxidation, carbonyl formation, 221
Cleaning, tapa cloth treatment, 172–173
Color, effect of deacidifying
 agents, 147,148–150*t*
Color change, measurement of degradation, 99
Conservation programs for books and paper,
 importance, 35
Coprosma species
 constituents, 196,197*t*
 source, 196

Cotton textiles, nondestructive evaluation
    of aging, 238–247
Crease removal, tapa cloth
    treatment, 173–174

D

Dactylopius coccus Costa
    constituents, 199
    source, 199
Damaging effects of exposure of paper to
    visible and near-UV radiation
chain breaking during extensive continuous
    exposure, 59,61f,62
characteristics of stock pulps, 55
development of hot-alkali-soluble matter
    in sheets of bleached pulp, 56,57f
    in sheets of filter paper, 59,60f
    in sheets of unbleached pulp, 56,57f
exposure to light sources, 55
measurements of properties, 55
percentage of links broken in bleached
    pulp, 56,58f,59
percentage of links broken in filter
    paper, 59,61f
thermal aging, 55
Deacidification, development, 13
Deacidification methods, advantages and
    disadvantages, 35–36
Degradation measurement, parameters, 99–100
Diethyl zinc deacidification process
    description, 21
    evaluation, 25,26t,27
    steps, 21
Diffuse-reflectance Fourier-transform IR
    spectroscopy, evaluation of textiles, 238
Dyeing properties of heat-aged linen,
    properties vs. dye structure, 166t,167

E

Energy dispersive X-ray spectrometry,
    analysis of silk flags, 134–142
Environmental conditions, effect on
    preservation of archival materials, 64

F

Fiber furnish, refining efficiency, 4
Fillers, choices for neutral–alkaline
    papermaking, 5–6
Filter paper
chain breaking during extensive continuous
    exposure, 59,61f,62
formation of hot-alkali-soluble matter
    during daylight exposure, 59,60f

Filter paper—*Continued*
    percentage of links broken during daylight
        exposure, 59,61f
Flax, use in textile constructions, 159
Fold endurance, papers of different types
    and ages, 38,41t

G

Galium verum L.
    constituents, 194–195
    source, 194
Gas deacidification, advantages and
    disadvantages, 20
Graft copolymerization
    application of other conservation
        procedures, 52
    basic method, 38
    commercial-scale considerations, 49
    container design, 45
    cost, 49
    dependence of fold endurance on weight
        gain for filter paper, 38,39f
    fold endurance for paper types and
        ages, 38,41t
    influencing factors, 41
    pressure vs. time plots, 45,48f
    processing stages, 49
    process scale-up steps, 41,45
    rate of polymerization vs. rate of radical
        formation, 38
    results from process scale up, 45,49
    schematic representation, 36,37f,38
    strength of γ-ray required vs. book
        height, 49,50f
    strength of γ-ray required vs. book
        mass, 49,51f
    temperature vs. time plots, 45,46–47f
    weight gain vs. radiation
        dose, 38,41,42–44f
    weight increase vs. fold endurance for
        mechanical paper, 38,40f

H

Heat-aged linen, dyeing
    properties, 166t,167
Heat-induced aging of linen
    direct dye structures, 161,164f
    dyeing procedure, 161,164f
    dyeing properties, 166t,167
    experimental procedure, 160
    heat-induced color changes, 160,161t,165
    materials, 160
    relative changes in tensile and abrasion
        characteristics, 160,161t,165

Heat-induced aging of linen—*Continued*
  scanning electron
    microscopy, 161,162–163*f*,165
  test methods, 160,161*t*,162–164*f*
  wide-angle X-ray scattering
    diffractograms, 161,164*f*,165
Historic silk flags, analysis for
  degradation factors, 134–142

I

Insect dyes, TLC, 208–216
Integrated complete book preservation
  program, discussion, 28–29
Interleaf vapor-phase deacidification
  process
  application, 23
  evaluation, 26*t*,28
Internal sizing
  application method, 3
  definition, 2
IR spectroscopy
  approaches to textile
    evaluation, 238–239,240*f*
  techniques, 237–238

K

Karabagh carpets, fragments, 209,211*f*,212
Kermococcus vermilio Planchon
  constituents, 200
  source, 199–200
Kerria lacca Kerr
  constituents, 201
  source, 200–201
Koppers process
  evaluation, 26*t*,28
  procedure, 22–23
Kubelka–Munk theory of diffuse reflectance
  calculation of parameters, 82
  derivation of equations, 82–83
  determination of color curves, 83
  experimental procedure for paper-yellowing
    study, 88
  factors affecting scattering
    coefficient, 86*t*,87
  Kubelka–Munk number calculation, 83
  paper-yellowing study, 87–88,89*f*
  physical interpretation of
    parameters, 83–84,85*f*
  problems and limitations, 83
  standard brightness calculation, 83

L

Lambert's cosine law, description of diffuse
  reflectance, 82

Linen
  heat-induced aging, 160–167
  natural aging, 159–160

M

Madder dyes
  constituents, 190,191–192*t*
  TLC, 201–208
Mass deacidification
  chemical characteristics, 20–23
  criteria, 24–25
  diethyl zinc deacidification process, 21
  evaluation of processes, 25,26*t*,27–28
  interleaf vapor-phase deacidification
    process, 23
  Koppers process, 22–23
  measurement of characteristics of
    treatments, 24
  recommendations for process
    selection, 29–30
  role in integrated complete book
    preservation program, 28–29
  Wei T'o nonaqueous book deacidification
    process, 22
Microfilming, preservation of books and
  paper, 34–35
Modern papers, fate, 13–14
Moisture regain, effect of deacidifying
  agents, 151,152*t*
Morinda citrifolia L.
  constituents, 195,197*t*
  source, 195
Morinda umbellata L., constituents, 195
Morpholine, use in gas deacidification, 20

N

Naturally aged cellulosic textiles, effect
  of alkaline deacidifying agents, 144–156
Neutral–alkaline method of internal sizing,
  advantages, 3
Neutral–alkaline papermaking
  economics, 4–11
  fiber furnish, 4
  fillers, 5–6
  process, 8–11
  sizing, 6,7*f*,8
Neutral–alkaline papermaking process
  draining on wire section, 9
  drying, 9–10
  effluent treatment, 10
  properties and price, 10–11
  reuse of broke and waste paper, 10
  size press, 10
  wet pressing, 9

Nonaqueous deacidification
advantages and disadvantages, 19
development, 19
neutralizers, 19
procedure, 19
Wei T'o process, 19–20
Nondestructive evaluation of aging in cotton
textiles
absorbance spectrum of unheated
cloth, 239,241f
development of carboxylate band, 239,244f
effect of aging time on
spectrum, 239,242–243f
effect of atmosphere on
spectrum, 239,243f
experimental procedure, 239
mirror background spectrum, 239,240f
morphology changes, 245
oxidation reactions of cellulose, 245
single-beam spectrum of unheated
cloth, 239,241f
summary of changes, 246
time vs. carboxylate OH stretching
absorbance, 239,244f

O

Oldenlandia umbellata L.
constituents, 195
source, 195
Optical analyses
definition, 81
types of interactions for paper, 81
Oxygen, deterioration of cellulose, 220–221

P

Paper
acid formation, 16–17
deacidification methods, 35–36
description, 2
effects of exposure to daylight
fluorescent lamplight, 54–62
Fresnel reflectivity, 82
hydrophilicity, 2
internal sizing, 2–3
invention, 14
permanence, 16
production, 2
world consumption, 14
Paper deterioration
acid-catalyzed hydrolysis, 35
oxidation, 35
sources of acid, 35

Papermaking
development of mechanical production
techniques, 15
history and development, 14–15
use of alum–rosin sizing, 15
Paper materials, use of optical tests in
characterization, 81
Paper yellowing
criterion of paper permanence, 87
effect of aging on scattering
coefficient, 87
optical analysis, 81
postcolor number, 87
use of Kubelka–Munk theory, 81–90
Parylene-C
advantages and disadvantages for use in
conservation of silk, 111
formation, 109,110f
properties, 109,111
Parylene-C conservation of silk
color change of films vs. light
exposure, 123,129f
color change vs. light exposure, 123,128f
effect of coating on historic silk
fabric, 131t,132
effect of coating thickness on tensile
properties, 114t
effect of light on breaking
load, 123,125f
effect of light on energy to
break, 123,127f
effect of light on strain to
break, 123,126f
effect of relative humidity on color
change, 123,124f
effect of relative humidity on tensile
properties, 120,121f
effect of time on breaking load, 115,116f
effect of time on color change, 120,121f
effect of time on energy to
break, 115,118f
effect of time on strain to
break, 115,117f
energy-to-break calculations, 113
hue angle of fabric vs.
temperature, 120,122f
light exposure, 123–128
light exposure procedure, 112–113
material preparation, 111
normalized color change of UV-irradiated
films, 123,130f,131
property measurements, 113
stress-to-break calculations, 113
tensile property degradation rates derived
from Arrhenius relationship, 119t,120
tensile property of coated and uncoated
fabric, 114t

Parylene-C conservation of silk—*Continued*
thermal degradation rate
constants, 115*t*,119
thermal exposure, 115–124
thermal exposure procedure, 111–112
Permanence of paper
determination by internal and external
factors, 16
evaluation, 16
Photoacoustic Fourier-transform IR
spectroscopy, evaluation of
textiles, 238,240*f*
Photothermal beam deflection
Fourier-transform IR spectroscopy,
evaluation of textiles, 238
Physical interpretation of Kubelka–Munk
parameters
absorbance of wood vs. refractive index of
liquid, 84,85*f*
absorption coefficient, 84
scattering coefficient, 84
surface reflectivity, 84
Porphyrophora hameli Brandt
constituents, 200
source, 200
Porphyrophora polonica L.
constituents, 200
source, 200
Postcolor number, definition, 87
Preservation technologies
graft copolymerization, 36–52
process criteria, 36

R

Red insect dyes
constituents, 199*t*
structures, 198*t*
Red madder dyes
advantages and disadvantages of HPLC, 190
advantages and disadvantages of TLC, 190
identification techniques, 189–190
importance, 188–189
Reflectance spectroscopy, theory, 82–83
Relative-humidity effects on accelerated
aging of paper
aging conditions, 65
effect of accelerated aging of loose
sheets and books, 66,76
effect of cycling relative humidity, 76–77
effect on brightness, 66,73–75*t*
effect on fold endurance of Kraft
books, 66,68*f*
effect on fold endurance of Kraft
sheets, 66,68*f*
effect on fold endurance of waterleaf
books, 66,67*f*

Relative-humidity effects on accelerated
aging of paper—*Continued*
effect on fold endurance of waterleaf
sheets, 66,67*f*
effect on increasing relative humidity, 66
effect on pH, 66,70–72*t*
effect on relative lifetime values, 66,69*t*
importance for understanding, 65
materials, 65
possible causes of degradation under
cycling relative humidity, 77–78
testing, 65
Relbunium ciliatum, constituents, 194
Relbunium hypocarpium, constituents, 194
Rubia akane, constituents, 194
Rubia cordifolia L.
constituents, 194
source, 193
Rubia peregrina L., constituents, 193
Rubia tinctorum L.
constituents, 193
source, 193

S

Scanning electron microscopy, heat-aged
linen, 161,162–163*f*,165
Scattering coefficients
description, 82
effect of aging, 87–88,89*f*
influencing factors, 86*t*,87
physical interpretation, 84
Screening of silk stabilizers
additives applied to preliminary
samples, 95,97*t*
amino nitrogen content of fabric vs.
stabilizer treatment, 104,105*t*
ammonia concentration of fabric vs.
stabilizer treatment, 103,104*t*
application of stabilizers, 98
color change of fabric vs. stabilizer
treatment, 102,103*t*
experimental procedure, 95–101
light-accelerated aging procedure, 100–101
objective, 94–95
parameters for degradation
measurement, 99–100
stabilizers applied to silk
fabrics, 95,97–98*t*
strength loss of fabric vs. stabilizer
treatment, 101*t*,102
thermally accelerated aging
procedure, 100
type, chemical name, and supplier of
stabilizer, 95,96*t*
types of stabilizers evaluated, 95
Sicilian coronation robe, identification of
insect dye, 215*f*,216

Silk
  conservation with Parylene-C, 108–132
  stabilization to light and heat, 94–105
Silk conservation techniques, 109
Silk fabric degradation in dark storage
  problems with standard test methods for
    assay of weighting materials, 134
  role of sulfuric acid, 134–135
Silk flag analysis
  ash and total sulfur analyses, 135
  ash content, 138,140t
  colorants, 138,141
  condition vs. manufacturing
    technology, 141
  description of flags,135,136t
  energy-dispersive X-ray spectrometric
    analysis, 135
  examples of flags, 135,137f
  future research, 141
  inorganic-weighting-agent content, 141
  pH, 138,140t
  pH and color analyses, 135
  qualitative energy-dispersive X-ray
    spectrometric results, 138,139t
  sulfur content, 138,140t
Sizing, reactions of alkyl ketene dimers and
  alkyl succinic anhydrides, 6,7f,8
Specular-reflectance Fourier-transform IR
    spectroscopy
  applications, 238
  use in nondestructive evaluation of aging
    in cotton textiles, 238–247
Stiffness, effect of deacidifying
    agents, 151,152t
Strength of paper, measurement, 36

T

Tapa cloth
  cultivation, 169
  decoration, 170
  factors influencing condition, 171
  harvesting, 169
  manufacture, 169–170
  source, 168–169
  uses, 170–171
Tapa cloth treatment
  adhesives, 175
  application of vacuum suction
    table, 176–178
  backing and tear repair, 174–175
  case studies, 179–184
  cleaning, 172–173
  crease removal, 173–174
  filling voids with free-hand cast paper
    pulp, 175–176
  treatment techniques, 172

Thin-layer chromatography of insect dyes
  comparison for Karabagh carpets, 212f
  dried insects, 209,210f
  preparation of samples, 208–209
  procedure, 209
Thin-layer chromatography of madder dyes
  Adlerdalmatika, 206,208
  comparison of extraction
    methods, 205,206f
  dyes of madder types, 203,204f
  dyes of plant similar to madder
    types, 203,204f,205
  improved extraction methods, 202–203
  preparation of samples, 201–202
  procedure, 203
Transmittance Fourier-transform IR
    spectroscopy, evaluation of textiles, 238
Tunicella
  identification of insect dye, 215f,216
  photograph, 214,215f

U

Unbleached pulp
  formation of hot-alkali-soluble matter
    during daylight exposure, 56,57f
  yellowing rates, 88,89f

V

Vacuum suction table
  case studies, 179–184
  formation of cast fills for filling
    voids, 178
  treatment of tapa cloth, 176–178
  use of environmental chamber, 177
  use of solvents to remove soils and
    stains, 178
  use of ultrasonic humidifier, 177–178
Ventilago mad(e)raspatana Gaertn.
  constituents, 198
  source, 197

W

Wei T'o nonaqueous book deacidification
    process
  description, 19–20
  evaluation, 26t,27
  procedure, 22
Wide-angle X-ray scattering, heat-aged
    linen, 161,164f,165

Y

Yellowing of paper, See Paper yellowing

Production: Donna Lucas
Indexing: Deborah H. Steiner
Acquisition: Cheryl Shanks

Elements typeset by Hot Type Ltd., Washington, DC
Printed and bound by Maple Press, York, PA

*Paper meets minimum requirements of American National Standard
for Information Sciences—Permanence of Paper for Printed Library
Materials, ANSI Z39.48–1984* ∞

Other ACS Books

Chemical Structure Software for Personal Computers
Edited by Daniel E. Meyer, Wendy A. Warr, and Richard A. Love
ACS Professional Reference Book; 107 pp;
clothbound, ISBN 0–8412–1538–3; paperback, ISBN 0–8412–1539–1

Personal Computers for Scientists: A Byte at a Time
By Glenn I. Ouchi
276 pp; clothbound, ISBN 0–8412–1000–4; paperback, ISBN 0–8412–1001–2

Biotechnology and Materials Science: Chemistry for the Future
Edited by Mary L. Good
160 pp; clothbound, ISBN 0–8412–1472–7; paperback, ISBN 0–8412–1473–5

Polymeric Materials: Chemistry for the Future
By Joseph Alper and Gordon L. Nelson
110 pp; clothbound, ISBN 0–8412–1622–3; paperback, ISBN 0–8412–1613–4

The Language of Biotechnology: A Dictionary of Terms
By John M. Walker and Michael Cox
ACS Professional Reference Book; 256 pp;
clothbound, ISBN 0–8412–1489–1; paperback, ISBN 0–8412–1490–5

Cancer: The Outlaw Cell, Second Edition
Edited by Richard E. LaFond
274 pp; clothbound, ISBN 0–8412–1419–0; paperback, ISBN 0–8412–1420–4

Practical Statistics for the Physical Sciences
By Larry L. Havlicek
ACS Professional Reference Book; 198 pp; clothbound; ISBN 0–8412–1453–0

The Basics of Technical Communicating
By B. Edward Cain
ACS Professional Reference Book; 198 pp;
clothbound, ISBN 0–8412–1451–4; paperback, ISBN 0–8412–1452–2

The ACS Style Guide: A Manual for Authors and Editors
Edited by Janet S. Dodd
264 pp; clothbound, ISBN 0–8412–0917–0; paperback, ISBN 0–8412–0943–X

Chemistry and Crime: From Sherlock Holmes to Today's Courtroom
Edited by Samuel M. Gerber
135 pp; clothbound, ISBN 0–8412–0784–4; paperback, ISBN 0–8412–0785–2

For further information and a free catalog of ACS books, contact:
American Chemical Society
Distribution Office, Department 225
1155 16th Street, NW, Washington, DC 20036
Telephone 800–227–5558